McGRAW-HILL NETWORKING AND TELECOMMUNICATIONS

Build Your Own

Trulove	*Build Your Own Wireless LAN* (with projects)

Crash Course

Louis	*Broadband Crash Course*
Vacca	*I-Mode Crash Course*
Louis	*M-Commerce Crash Course*
Shepard	*Telecom Convergence, 2/e*
Shepard	*Telecom Crash Course*
Bedell	*Wireless Crash Course*
Kikta/Fisher/Courtney	*Wireless Internet Crash Course*

Demystified

Harte/Levine/Kikta	*3G Wireless Demystified*
LaRocca	*802.11 Demystified*
Muller	*Bluetooth Demystified*
Evans	*CEBus Demystified*
Bayer	*Computer Telephony Demystified*
Hershey	*Cryptography Demystified*
Taylor	*DVD Demystified*
Bates	*GPRS Demystified*
Symes	*MPEG-4 Demystified*
Camarillo	*SIP Demystified*
Shepard	*SONET / SDH Demystified*
Topic	*Streaming Media Demystified*
Symes	*Video Compression Demystified*
Shepard	*Videoconferencing Demystified*
Bhola	*Wireless LANs Demystified*

Developer Guides

Guthery	*Mobile Application Development with SMS*
Richard	*Service and Device Discovery: Protocols and Programming*

Professional Telecom

Smith/Collins	*3G Wireless Networks*
Bates	*Broadband Telecom Handbook, 2/e*
Collins	*Carrier Grade Voice over IP*
Harte	*Delivering xDSL*
Held	*Deploying Optical Networking Components*
Minoli/Johnson/Minoli	*Ethernet-Based Metro Area Networks*
Benner	*Fibre Channel for SANs*
Bates	*GPRS*

Sulkin	*Implementing the IP-PBX*
Lee	*Lee's Essentials of Wireless*
Bates	*Optical Switching and Networking Handbook*
Wetteroth	*OSI Reference Model for Telecommunications*
Russell	*Signaling System #7, 4/e*
Minoli/Johnson/Minoli	*SONET-Based Metro Area Networks*
Nagar	*Telecom Service Rollouts*
Louis	*Telecommunications Internetworking*
Russell	*Telecommunications Protocols, 2/e*
Minoli	*Voice over MPLS*
Karim/Sarraf	*W-CDMA and cdma2000 for 3G Mobile Networks*
Bates	*Wireless Broadband Handbook*
Faigen	*Wireless Data for the Enterprise*

Reference

Muller	*Desktop Encyclopedia of Telecommunications, 3/e*
Botto	*Encyclopedia of Wireless Telecommunications*
Clayton	*McGraw-Hill Illustrated Telecom Dictionary, 3/e*
Radcom	*Telecom Protocol Finder*
Pecar	*Telecommunications Factbook, 2/e*
Russell	*Telecommunications Pocket Reference*
Kobb	*Wireless Spectrum Finder*
Smith	*Wireless Telecom FAQs*

Security

Nichols	*Wireless Security*

Telecom Engineering

Smith/Gervelis	*Cellular System Design and Optimization*
Rohde/Whitaker	*Communications Receivers, 3/e*
Sayre	*Complete Wireless Design*
OSA	*Fiber Optics Handbook*
Lee	*Mobile Cellular Telecommunications, 2/e*
Bates	*Optimizing Voice in ATM / IP Mobile Networks*
Roddy	*Satellite Communications, 3/e*
Simon	*Spread Spectrum Communications Handbook*
Snyder	*Wireless Telecommunications Networking with ANSI-41, 2/e*

BICSI

Network Design Basics for Cabling Professionals
Networking Technologies for Cabling Professionals
Residential Network Cabling
Telecommunications Cabling Installation

GIGABIT ETHERNET FOR
METRO AREA NETWORKS

Gigabit Ethernet for Metro Area Networks

Paul Bedell

McGraw-Hill
New York   Chicago   San Francisco   Lisbon
London   Madrid   Mexico City   Milan   New Delhi
San Juan   Seoul   Singapore   Sydney   Toronto

The McGraw-Hill Companies

Cataloging-in-Publication Data is on file with the Library of Congress.

1 2 3 4 5 6 7 8 9 0   DOC/DOC   0 9 8 7 6 5 4 3 2

ISBN 0-07-139389-7

*The sponsoring editor for this book was Steve Chapman and the production supervisor
was Sherri Souffrance. It was set in Century Schoolbook by MacAllister Publishing
Services, LLC.*

Printed and bound by RR Donnelley.

To Paula, Aaron, Ryan, Bob, and to God

CONTENTS

Contents

ACKNOWLEDGMENTS

I'd like to extend my utmost appreciation to the following friends, family, professional associates, coworkers, and acquaintances. Some of these individuals simply kept me in the loop on Ethernet news. Some sent me articles that became topics or information for this book. Others just gave me advice and assistance. They all took part in the development of this effort in some way.

So thanks to Bill Sudbrook, Chad Hiestand, Theresa Elliott, Rick Tonielli, Carissa Hernandez, Toya Drake, Linda Morgan, Tom Grace, Bob Walters, Ruth Anne Renaud, Patsy Torbert, Steve Bedell, Julie Kobach, Dan Palicka, Carol Huss, and Vish Ramamurti. Special thanks also goes out to Mark K. Curtis for his contributions on case studies, and another McGraw-Hill author and friend Steven Shepard for his guidance when I needed it.

Thanks to my neighbors Don and Sue Vlasaty for their undying support during the entire year it took me to write this book.

I'd also like to thank the following people from professional research organizations who graciously gave me permission to use information from their reports in this book: Fred Knight of *Business Communications Review Magazine*, Julie Brandt of the International Engineering Consortium (IEC), Sam Lucero (Cahners In-Stat), Daryl Schoolar of In-Stat/MDR, Ted Smith and Evan Moltz of TechGuide.Com (Cnet), Nick Maynard of the Yankee Group, Rochelle Barnich of New Paradigm Resources Group, Tim Luke of Lehman Brothers, Norm Bogen of In-Stat/MDR.

Thanks to my wife Paula, for her patience and understanding while I toiled on this project for an entire year . . . at night, on weekends, whenever.

Last but not least, I'd like to extend my gratitude to Steve Chapman, Editor-in-Chief of McGraw-Hill Professional. As always, Steve was an effective guide as I worked to develop, begin, and complete this book.

PREFACE

One of the most dynamic industries on the face of the earth is the telecommunications industry. Today, telecom describes the need and ability to move all types of information across vast expanses of geography using many types of media and protocols. Boxes (servers, various types of switches, and routers), circuits, and the Internet are transforming the way we live and work. A fundamental component of this communication in the networked world is Ethernet, the network protocol that defines how all networked business- and Internet-related communications begin and end.

This book explains how Ethernet technology is moving out of the *local area network* (LAN) of the twenty-first century into *metropolitan area network* (MAN) and *wide area network* (WAN) environments. It is presumed that the reader of this book has a basic understanding of networking and has worked in the computer or telecommunications industry for three to five years. *Gigabit Ethernet for Metro Area Networks* is aimed at the enterprise or carrier network analyst, manager, or executive who needs to know what is happening with Ethernet today and what he or she can do to capitalize on the changes that are occurring in this area. The purpose of this book is to educate the reader so informed decisions can be made regarding strategic network management.

The traditional view of Ethernet is changing dramatically. It is extending its capability out of the LAN and into the MAN at a rapid pace. Entire new breeds of carriers have predicated their entire existence on supplying the means to offer MAN-based Ethernet services. Equipment platforms are now available to mesh Ethernet with legacy networks and protocols (such as SONET). Additionally, Ethernet will soon likely be capable of managing end-to-end transmissions in the WAN environment, possibly in native format from end to end! *Gigabit Ethernet for Metro Area Networks* explains this evolution in detail. It discusses the basics of Ethernet, what technologies can complement Ethernet in a metro area environment, and what technologies Ethernet will be competing against in a MAN. The players are described in this book, including equipment suppliers and service providers. The business perspective is also examined, exploring the ways in which an all-Ethernet MAN can be justified economically as a positive migration for any company to make when multiple locations are involved. Case studies are included to bring a dose of reality to the business equation.

It's important to remember that this book is not meant to be an engineering tome. The objective is to let the reader know what's going on with the Ethernet revolution, what's driving it, how Ethernet can be used, and why this is important..

Like my previous books, I've done my best to discuss technical concepts in plain, easy-to-understand terms. It is my sincere hope that *Gigabit Ethernet for Metro Area Networks* enables the reader to understand the impact that Ethernet is having on today's telecommunications and computer networks. Readers of this book who are network managers or IT executives should then be able to make intelligent choices when it comes time to modify or redesign the networks they manage. Readers who are just seeking a more thorough understanding of metro area Ethernet should sufficiently expand their knowledge upon finishing this book.

—Paul Bedell

INTRODUCTION

Gigabit Ethernet for Metro Area Networks discusses the public network evolution that is now under way in the United States, where many enterprises and carriers alike are modifying their approach to *metropolitan area network* (MAN) design and management. The move to Ethernet technology to handle MAN transport is well under way and promises to continue growing for years to come.

This book is organized in a manner that leads the reader through this change in a logical, building-block manner. It's presumed that the reader has a basic understanding of the data networking fundamentals, the protocols that can be used in data transport, and the various means of transport available through the *Public Switched Telephone Network* (PSTN).

The perspectives of enterprise and carrier networks will be presented. The intention is to compartmentalize these discussions whenever possible. But the truth of the matter is that the very nature of this paradigm shift mandates that both the enterprise network and the carrier network infrastructure undergo this expanded use of Ethernet simultaneously, at least to some degree. How else could an enterprise migrate to an Ethernet-only topology to connect multiple metro locations together if the carriers in a metro area don't also have a medium-haul Ethernet capability in their own networks? If that were not the case, the seamless transport of enterprise traffic across a metro area would have to be encapsulated into *wide area network* (WAN) protocols such as *Asynchronous Transfer Mode* (ATM) or frame relay, which defeats the entire purpose of migrating to Ethernet in the metro. The objective is to seamlessly transport Ethernet-based *local area network* (LAN) traffic across an urban area. Therefore, when it makes sense, discussions on Ethernet in both the enterprise and carrier environment will sometimes be meshed (sorry for the pun).

Chapter 1, "Fundamentals of Ethernet Technology," reviews the origins and basics of Ethernet technology. Included in this review are the basics of the 802.3 standard—*Carrier Sense Multiple Access with Collision Detection* (CSMA/CD) technology. This chapter also discusses Ethernet's place in the LAN and why it has become the de facto standard in today's LAN. A brief review of the Token Ring standard is also included, which is accompanied by an explanation of why Ethernet became more popular than Token Ring over the years and why Ethernet now dominates over 95 percent of all corporate LAN environments. The nature of Ethernet transmissions in the LAN are reviewed and descriptions of *network interface cards* (NICs), the LAN switch, and the LAN server are provided. Finally, this chapter pulls it all together by briefly delivering a history of the Ethernet hub, the migration to Ethernet switching in the 1990s, and the development of *virtual LAN* (VLAN) technology.

Chapter 2, "Metro Area Ethernet," delivers an overview of MANs, today's MAN marketplace, and what's driving the rapid deployment of optical (Gigabit) Ethernet in MANs. Various applications are discussed, and a detailed overview of the recently ratified 10 GigE standard is also included. This chapter covers Ethernet access, and dedicates an entire section to Ethernet in the first mile and the rapidly evolving *Ethernet Passive Optical Networks* (EPONs) technology.

Chapter 3, "The Metro GigE Marketplace," reviews the metro area Ethernet marketplace. The service providers and their related Ethernet offerings are discussed, and a section on the key players in the equipment space and the hot developments in that arena is provided. The equipment marketplace is becoming very competitive as Ethernet rapidly makes its move out of the LAN and into the MAN, and soon the WAN.

The service provider marketplace includes many different types of companies making a run at providing Ethernet transport services in the metro area. Most of these new carriers are struggling under the weight of the telecom implosion of 2001 and 2002. Some carriers are offering very disruptive pricing schemes, which sometimes have very catchy—but strange—names. However, like many of the dead *Competitive Local Exchange Carriers* (CLECs), some will not be available by 2003. Others are very nontraditional (such as cable TV companies, municipalities, and utilities), and their approach and service offerings will also be explored.

Chapter 4, "Competing Technologies," contains a review of current technologies that will compete with Ethernet for a share of the metro area transport marketplace. This mostly includes legacy technologies based on *time division multiplexing* (TDM) and legacy packet transport technologies such as frame relay and ATM.

Chapter 5, "Complementary Technologies and Protocols," discusses the nature of the many technologies that will complement metro area Ethernet transport, including *Resilient Packet Ring* (RPR), next-generation SONET, and *dense wave division multiplexing* (DWDM).

Upon completing this book, the reader should walk away with a comprehensive understanding of the place of Ethernet in the metropolitan networking world of the twenty-first century. The key takeaway is how the increasing reliance on Ethernet technology is changing the way businesses and carriers are moving data and how this renewed focus on Ethernet is changing the way networks are designed. The applications that can be sent over these types of networks may have a major impact on networking itself and how we live. The simplicity of Ethernet and its interworking with developing technologies will largely define the face of telecommunications for our economy and way of life well into the early twenty-first century.

Fundamentals of Ethernet Technology

As indicated in the Preface, it is assumed that the reader has three to five years' experience in the computer or telecommunications field. However, in the interest of ensuring that a common understanding of Ethernet is fostered at the beginning of this book, a review of Ethernet fundamentals, specifications, and hardware is delivered in this chapter. If nothing else, a refresher in the technology that this book revolves around certainly wouldn't hurt. That said, let's move forward.

1.1   The Invention of Ethernet and the 802.3 Standard

Dr. Robert M. Metcalfe invented Ethernet technology at the Xerox Palo Alto Research Center in the 1970s. It was designed to support research on the "office of the future," which included one of the world's first personal workstations—the Xerox Alto. The first Ethernet system ran at approximately 3 *megabits per second* (Mbps) and was known as *experimental Ethernet*.

Key: The *ether* in Ethernet indicates it can use any physical medium to transport information and communicate: copper, optical fiber, and even air (wireless in the form of radio or laser).

It should be noted that today the term *Ethernet system* can refer to a *local area network* (LAN) environment, a *metropolitan area network* (MAN) environment, and soon a *wide area network* (WAN) environment.

Formal specifications for Ethernet were published in 1980 by the multivendor consortium that created the *DEC-Intel-Xerox* (DIX) standard. Digital, Intel, and Xerox developed the technology jointly after working together to create DIX. This effort converted experimental Ethernet into an open (interoperable), production-quality Ethernet system that operated at 10 Mbps in a LAN environment.

A LAN is a group of computers and associated devices (printers) that share a common communications line and typically share the resources of a single processor or server within a small geographic area (such as an office building). Usually, the server has applications and data storage that are also shared in common by multiple computer users. A LAN may serve as few as two or three users (a home network) or as many as thousands of users in a *multitenant unit* (MTU) such as a high-rise office building.

Ethernet technology was then adopted for standardization by the LAN standards committee of the *Institute of Electrical and Electronics Engineers* (IEEE 802). The IEEE describes itself as "the world's largest technical professional society—promoting the development and application of electrotechnology and allied sciences for the benefit of humanity, the advancement of the profession, and the well being of our members." The IEEE fosters the development of network standards that often become national and international standards. The organization publishes a number of journals, and has many local chapters and several large societies in special areas, such as the IEEE Computer Society. The IEEE Ethernet standard was published in 1985 and was given the formal title IEEE 802.3 *Carrier Sense Multiple Access with Collision Detection* (CSMA/CD) Access Method and Physical Layer Specifications. Since then, virtually all computer vendors have strongly embraced Ethernet. The IEEE standard was also adopted by the *International Telecommunications Union* (ITU), which made Ethernet a worldwide networking standard. To underscore the degree to which the industry has embraced Ethernet, consider several facts:

- In 1999, 98 percent of all LAN ports shipped were Ethernet.
- In 2000, 182,000,000 Ethernet ports shipped.
- As of 2001, there are over 300,000,000 Ethernet-based computers in the world.

Although several networking options are available for LANs, Ethernet remains the most common LAN technology in use today. The IEEE 802.3 standard defines Ethernet standards for LAN configuration (topology), node networking (access method), and media use (speeds and feeds).

Ethernet has emerged as the most common LAN technology for a variety of reasons, but the most basic reasons include

- Low cost
- Easy installation and provisioning
- Multiple topology options
- A variety of basic communication and networking needs

Key: The widespread popularity of Ethernet ensures that there is a large market for Ethernet equipment, which also helps keep the technology competitively priced.

1.2 Ethernet Media

The *American National Standards Institute / Electronic Industries Association* (ANSI/EIA) Standard 568 is one of several standards that specify categories (commonly referred to as CAT) of twisted-pair cabling systems in terms of the data rates they can sustain. The specifications describe the cable material used as well as the types of connectors and junction blocks that will be used in order to conform to a specific category. These categories are described in Table 1-1.

Although longer connections for GigE use optical fiber, the goal in most organizations is to leverage the CAT 5 twisted-pair wiring they already have in place for connections out to the desktop. Four pairs of twisted-pair wires are commonly used in this context. The wires are all housed in one unshielded cable sheath. *Unshielded* means that there is no special metal insulation on the inside of the cable sheath to protect the transmissions from interference due to the proximity of things such as electric motors or fluorescent lighting.

Table 1-1

Cabling
classifications

Category	Maximum Data Rate	Typical Application
CAT 1	Less than 1 Mbps	Analog voice *Plain Old Telephone Service* (POTS) *Integrated Service Digital Network* (ISDN) *Basic Rate Interface* (BRI) Doorbell wiring
CAT 2	4 Mbps	Mainly used in the IBM cabling system for Token Ring networks
CAT 3	16 Mbps	Voice and data on 10BaseT Ethernet
CAT 4	20 Mbps	Used in 16 Mbps Token Ring, otherwise not used much
CAT 5	100 Mbps 1,000 Mbps (4 pairs)	100 Mbps *copper distributed data interface* (CDDI) 155 Mbps *Asynchronous Transfer Mode* (ATM) *Gigabit Ethernet* (GigE) (in building only)
CAT 5e	100 Mbps	100 Mbps CDDI 155 Mbps ATM
CAT 6	200 to 250 Mbps	Super-fast broadband applications

Key: Although CAT 3 and CAT 5 cables may look identical, CAT 3 is tested to a lower set of specifications and can cause transmission errors if pushed to faster speeds. CAT 3 cabling is near-end crosstalk certified for only a 16 MHz signal, whereas CAT 5 cable must pass a test to ensure it can transmit at 100 Mbps.

1.2.1   Ethernet Media Classifications

Ethernet has drawn increased attention in recent years as speeds have improved from the standard 10 Mbps to faster rates of 100 Mbps (Fast Ethernet), 1 *gigabit per second* (Gbps) (GigE), and now 10 Gbps (10 GigE). One Gbps equates to 1 billion bits per second. The 10 GigE standard was approved and published in June 2002.

A variety of media is available to meet the transmission needs of an Ethernet network. Even though the cabling of a network may vary, the other specifications remain the same. In the interests of historical accuracy, even though some of the media specifications listed in the following section are only used in legacy systems today—or not at all—they are listed to give a sense of how Ethernet technology has evolved over the past 20 years.

The number 10 in the following standards stands for 10 Mbps, the original Ethernet speed. The word *Base* in the following definitions stands for *baseband*. Baseband describes a telecommunication system in which information is carried in digital form on a single unmultiplexed signal channel over the transmission medium. Ethernet is a baseband system that uses *Manchester encoding* of high and low voltages to place bits (zeros and ones) on a wire pair.

Key: In Manchester encoding, each bit time contains a transition in the middle of the bit transmission. A transition from low to high represents a 0 bit and a transition from high to low represents a 1 bit. In most of the following cases, the transmission medium is CAT 5 twisted-pair copper cable.

10Base5   Also known as *thick coax* or *Thicknet*, this was one of the original Ethernet cabling solutions. It allowed cable runs up to a maximum length of 500 meters. Clamp-like vampire taps connected nodes to the central cable. Above-average cost and large cable size led to its declining use. The 5 in 10Base5 stands for 500 meters.

■ **10Base2**   Also known as *thin coax* or *Thinnet*, this was another early Ethernet cabling option. It allowed maximum cable runs of just under 200 meters. *British Naval Connectors* (BNCs) linked users to the network and bayonet caps terminated the signals. The biggest disadvantage of 10Base2 technology is that if cabling becomes defective, the whole network goes down. The 2 in 10Base2 stands for 200 meters.

■ **10BaseT**   This uses CAT 3, CAT 5, CAT 5e, or CAT 6 *unshielded twisted-pair* (UTP) copper cable and RJ-45 connectors (low-cost phone wire and jacks) to connect users to an Ethernet hub or switch in a wire closet in a star pattern. See Figure 1-1 for an illustration of a typical LAN layout from cubes and offices to the wiring closet.

■ The hub or switch in the wiring closet is known as the *hubbing point*. The terms *node, personal computer* (PC), and *workstation* will be synonymous from this point forward. Although a failure of a hub or switch would be a single point of failure, 10BaseT networks are also easy to troubleshoot. 10BaseT is the most popular media option to connect workstations to Ethernet hubbing points today. There is a

Figure 1-1
CAT 5 connections from cubes and offices to wiring closet

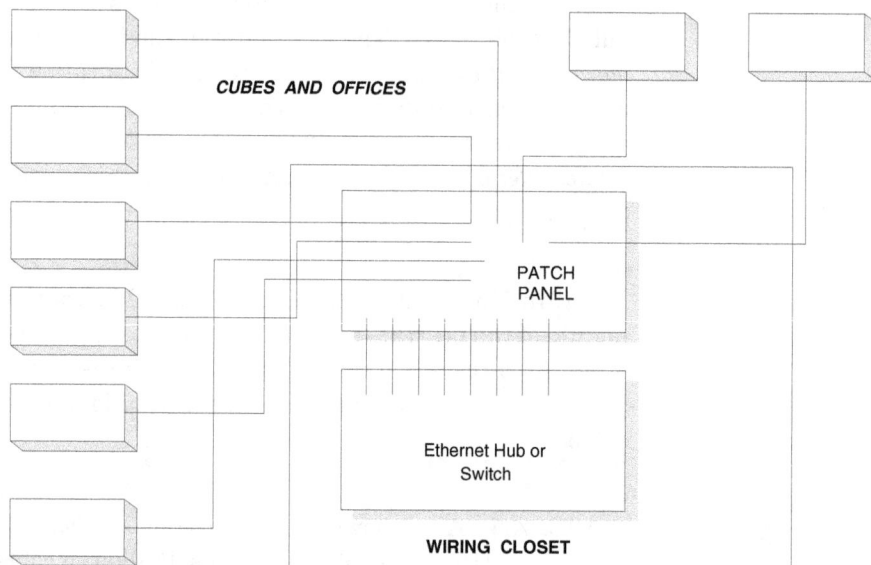

massive embedded base of 10BaseT throughout the world; however, many enterprises have migrated to 100BaseT (also known as Fast Ethernet) as bandwidth requirements increase. The *T* in 10BaseT stands for *twisted pair*.

10BaseF   This uses optical fiber and light pulses to transmit information instead of electrical current. Similar to 10BaseT, users attach to a central repeater or concentrator (hubbing point) in a star pattern. Network segments—the span between an Ethernet hub or switch and a workstation—can be up to 2,000 meters long due to the increased reach that is available with fiber optics, making it a great choice for network backbones. A network backbone is the part of the network that all hubbing or switch points connect to in order to access central file storage devices (servers) or the Internet. Think of vertical cabling in a high-rise building—that could be considered the network backbone. The *F* in 10BaseF stands for *fiber*.

100BaseT and 100BaseF   These are the media options for Fast Ethernet (IEEE 802.3u standard). They use UTP copper cable and optical fiber, respectively, to transmit at 100 Mbps. In 1995, the IEEE officially adopted the 802.3u 100BaseT Fast Ethernet standard as the successor to 10BaseT.

1000BaseT, **1000BaseLX, and 1000BaseSX**   These are the three types of media used for GigE (IEEE 802.3z) networks. UTP, *single-mode fiber* (SMF), and *multimode fiber* (MMF) cabling allow for transmission rates of 1,000 Mbps (also known as 1 Gbps) at a variety of distances and are typically found in the network backbone within a building. 1000BaseLX is also used in a MAN environment to connect LANs across metropolitan areas. The permissible end-to-end distance between endpoints in a 1000Base*n* network depends on the type of vendor equipment used at each end and whether repeaters are used somewhere in the middle of the signal path in the service provider's network.

The physical distance limitations of Ethernet media may be further restricted due to the timing constraints imposed by the collision detection aspect of CSMA/CD. The exception to this rule is GigE in the metro area, which operates in full-duplex mode—CSMA/CD is not used. This concept will be discussed in detail in Chapter 2, "Metro Area Ethernet." Table 1-2 depicts the types of Ethernet transmission, the medium used, and the maximum transmission distance allowed on each particular medium.

Table 1-2

Ethernet media summary

Type	Medium	Max Length	Connector
10Base5	Coax	500 m	Vampire tap
10Base2	Coax	185 m	BNC
10BaseT	UTP	100 m	RJ-45
10BaseF	Fiber	2,000 m	ST/SC/MT-RJ-45
100BaseT	UTP	100 m	RJ-45
100BaseF	Fiber	2,000 m	ST/SC/MT-RJ-45
1000BaseT	UTP	100 m	RJ-45
1000BaseLX	SMF	5,000 m	ST/SC/MT-RJ-45
1000BaseSX	MMF	550 m	ST/SC/MT-RJ-45

1.3 The Ethernet Standard

The Ethernet standard consists of three basic elements:

- The physical medium used to carry Ethernet signals between computers.

- A set of *Medium Access Control* (MAC) rules embedded in each Ethernet *network interface adapter* (also known as a *network interface card* [NIC]) that enable multiple computers to fairly arbitrate access to the shared Ethernet channel. These rules are programmed into the firmware of every Ethernet NIC.

Key: An NIC is an adapter card that physically connects a computer to a network cable. It is an inherent part of any PC in a business environment. The terms *network adapter*, *Ethernet adapter*, and *NIC* are all synonymous. In this book, the term NIC will be used.

- An Ethernet frame that consists of a standardized set of bits (ones and zeros) used to carry data over the system.

Each Ethernet-equipped computer, such as a PC workstation or network server, operates independently of all other stations on the network—there is no central controller. All workstations attached to an Ethernet network are connected to a shared signaling system (the medium). Ethernet signals are transmitted serially, one bit at a time, over the shared signal channel to all attached workstations—they are broadcasted. To send data, a workstation first listens to the channel, and when the channel is idle, the station transmits its data in the form of Ethernet frames. After each frame transmission, all workstations on the network must contend equally for the next opportunity to transmit a frame. This ensures that access to the network channel is fair and that no single workstation can lock out the other workstations by monopolizing access to the medium (channel) for a certain period of time. Access to the shared channel is determined by the MAC mechanism embedded in the Ethernet NIC located within each workstation or server.

Key: In today's Ethernet networks, all devices (such as PC workstations and network servers) are connected to the LAN access cable, such as CAT 5 UTP, CAT 6 UTP, and, in rare cases, coaxial cable. These devices then compete for access to the network using the protocol CSMA/CD. CSMA/CD is the MAC referred to previously as an element of Ethernet systems. It will be discussed in a subsequent section.

1.3.1   Ethernet Frames and Transmission

The heart of the Ethernet system is the Ethernet frame, which is used to deliver data between computers. The frame consists of a set of bits organized into several fields. These fields include address fields, a variable-size data field that carries 46 to 1,500 bytes of payload data, and an error-checking field that checks the integrity of the bits in the frame to make sure that the frame has arrived intact.

Key: All of the logic implementing the Ethernet protocol is implemented in the Ethernet NIC. Ethernet payload data is encapsulated in a fixed packet format. The data field must be at least 46 bytes and no more than 1,500 bytes (with a format-framing overhead of 36 and 1.7 percent, respectively).

The sections of a standard Ethernet frame are described in the following list:

- Ethernet nodes use the *preamble* field to synchronize with the signal frequency transmitted over the LAN. The preamble bit pattern is the same for all Ethernet media types except Fast Ethernet.

  The first two fields in the Ethernet frame carry 48-bit addresses called the *destination* and *source addresses*. To maintain uniformity through all Ethernet networks everywhere, the IEEE controls the assignment of these addresses by administering a portion of the address field. The IEEE does this by providing 24-bit identifiers called *organizationally unique identifiers* (OUIs). A unique 24-bit identifier is assigned to each company that builds Ethernet NICs. Examples of such companies include 3Com, IBM, Intel, and Xircom. The companies that manufacture Ethernet NICs, in turn, create 48-bit addresses using the IEEE-assigned OUI as the first 24 bits of the address. This 48-bit address is also known by any one of three names: the physical address, hardware address, or MAC address. This unique 48-bit address is preassigned to each Ethernet NIC when it is manufactured, which simplifies the setup and operation of the network significantly. One key advantage to having preassigned addresses is that it prevents network administrators from getting involved in administering the addresses for different groups using the network. This could add an unwelcome burden to the already challenging job of administering networks.

- The *destination address* field identifies the receiving station(s) by the physical MAC address, logical network address (*Transmission Control Protocol / Internet Protocol* [TCP/IP] address), and/or broadcast/multicast address.

- The *source address* field also identifies the sending station by the physical MAC address, logical network address, and/or broadcast/multicast address.

- The *type / length* field specifies either the upper-layer protocol ID (for TCP/IP packets) or data length indicator.

- The *data* field contains upper-layer user data. Because Ethernet frames must always be at least 64 bytes in length, the data field is padded if the user data is less than 64 bytes. Padding is also referred to as *bit stuffing*.

- The *Frame Check Sequence* (FCS) uses a checksum for error control purposes. A checksum is a validation algorithm that uses binary division. Packets with invalid checksums are discarded. However, no error message is sent to the source node because Ethernet is a layer 2 technology in the *Open Systems Interconnect* (OSI) model. Higher layers manage error detection and correction, and retransmissions when they're required.

This standard format enables Ethernet packets to be easily mapped, bridged, or routed into many other network protocols, including TCP/IP, ATM, and frame relay. See Figure 1-2 for an illustration of a standard Ethernet frame.

As each Ethernet frame is sent to the shared signal channel in a LAN, *all* Ethernet interface cards at *all* workstations look at the first 48-bit field of the frame, which contains the destination address of the intended recipient of the Ethernet transmission. The interface cards of all workstations compare the destination address of the frame with their own MAC address. The Ethernet NIC (workstation) with the same address as the destination address in the transmitted frame will read in the entire frame and deliver it to the networking software running on that computer. All the other NICs at all the other workstations will stop reading the frame when they discover that the destination address does not match their own MAC address.

1.4 Basic Components and Operations

Ethernet LAN technology transmits information between computers at speeds of 10 million, 100 million, 1 billion bits per second ($\approx$1,000 Mbps),

Figure 1-2
The standard
Ethernet frame

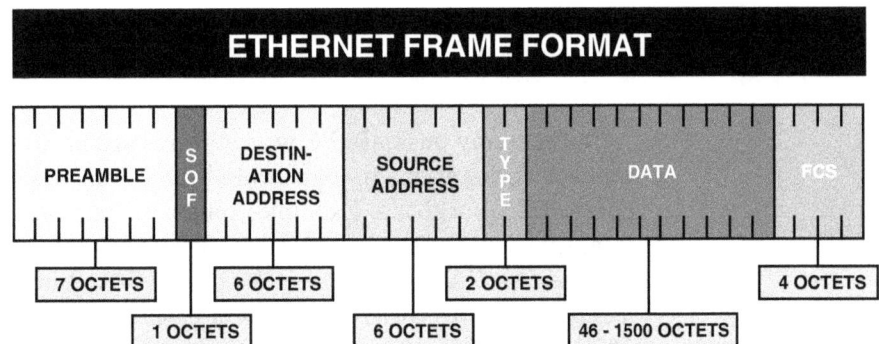

ETHERNET FRAME FORMAT						
PREAMBLE	SOF	DESTIN-ATION ADDRESS	SOURCE ADDRESS	TYPE	DATA	FCS
7 OCTETS	1 OCTETS	6 OCTETS	6 OCTETS	2 OCTETS	46 - 1500 OCTETS	4 OCTETS

and now 10 billion bits per second (10 GigE). Currently, the most widely used version of Ethernet technology is the 10 Mbps twisted-pair cable variety, but many LAN managers are now migrating key segments of their LANs to 100 Mbps and even 1,000 Mbps (GigE). The most recent Ethernet standard—GigE—defines a system that operates over copper cable (in building) as well as optical fiber media. Now that the 10 GigE standards are complete, they will require optical fiber as their transport media, for any scenario.

10 Mbps Ethernet LANs (10BaseT) can use thin coaxial (this is rare today); CAT 3, CAT 5, CAT 5e, or CAT 6 UTP copper cable; and fiber-optic systems. Fast Ethernet, also known as 100BaseT, provides transmission speeds up to 100 Mbps, and today is most commonly used for LAN backbone systems. In this type of application, Fast Ethernet will support traffic from workstations with 10BaseT NICs and back haul the traffic to server farms. For example, 10 Mbps workstations would connect to a 10/100 Mbps Fast Ethernet switch in a wire closet on a given floor of a building. Then, Fast Ethernet would be used to route traffic through the backbone (up building risers) and to connect other Fast Ethernet switches on other floors of the building together. All of those backbone Fast Ethernet switches would then have their traffic terminated into servers and/or a high-speed Ethernet switch/router at the main data or server room.

Key: Corporate LAN backbones must always operate at an Ethernet speed that's incrementally higher than the highest-speed NICs operating in the user environment. For example, if all end users have 10 Mbps NICs in their workstations, then it's logical to use 100 Mbps speeds in the LAN backbone to aggregate and route 10 Mbps traffic. If a sufficient number of end users whose workstations are operating at 100 Mbps speeds is present, then GigE (1,000 Mbps) speed would be required in the LAN backbone to aggregate and route the LAN traffic. This architecture serves to avoid creating a network bottleneck (choke point). Figure 1-3 depicts this configuration.

Today, many business PC workstations are now using Fast Ethernet to connect workstations that generate high volumes of data traffic to the corporate LAN backbone. Examples might include people who use graphics-intensive applications as a regular part of their job, such as *computer-aided engineering* (CAE), *computer-aided design* (CAD), or *computer-aided manufacturing* (CAM). Other examples of desktop workstations that might use 100 Mbps Ethernet connections are computer programmers who have a

Figure 1-3
100 Mbps Fast
Ethernet collapsed
backbone LAN
topology

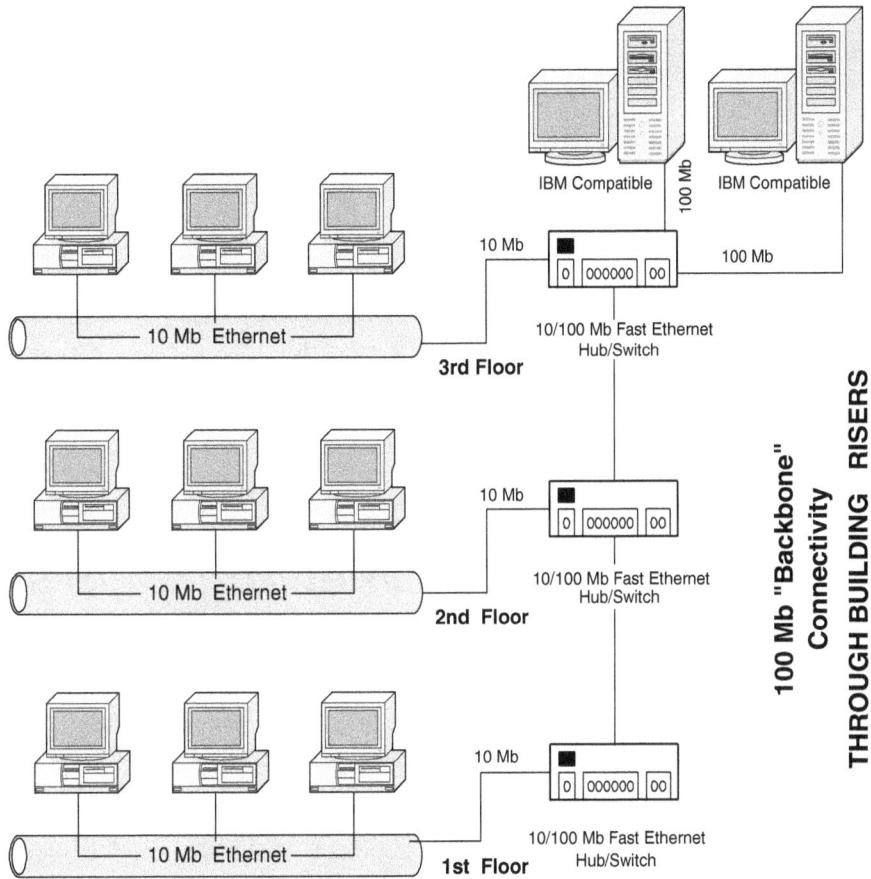

need to upload large files for compiling or people in the medical field working with large, graphics-intensive medical imaging systems. In these scenarios, it's likely that GigE would be used in the backbone itself in order to aggregate the traffic from many workstations operating at Fast Ethernet speeds and avoid creating a bottleneck.

From the time of the first Ethernet standard, the specifications and the rights to build Ethernet technology have been made easily available to anyone. This openness, combined with the ease of use and robustness of the Ethernet system, resulted in a large Ethernet market. This is another reason why Ethernet is so widely implemented in the computer industry, and, by default, the business world.

The vast majority of computer and server vendors today equip their products with 10 Mbps Ethernet attachments, making it possible to link all types of computers with an Ethernet LAN. As the 100 Mbps standard has become more widely adopted, desktop and laptop computers are now equipped with dual-mode, autosensing Ethernet adapters that may operate at both 10 and 100 Mbps, depending on the network environment.

Key: The ability to link a wide range of computers using a vendor-neutral, open network technology such as Ethernet is an essential feature for today's network managers.

Modern LANs must support a wide variety of computers purchased from different vendors. This requires a high degree of transparent network interoperability of the kind that Ethernet provides.

1.5   Ethernet Topologies

Ethernet networks can use a number of topologies, the most common being the bus, star, and tree layouts. To understand how signals flow over the set of media segments that make up an Ethernet system, it helps to understand the topology of the system.

Key: The signal topology of the Ethernet is also known as the logical topology to distinguish it from the actual physical layout of the media cables. This applies whether the network is in a small business office in a strip mall or a large multitenant, 30-story office building.

The logical topology of an Ethernet network provides a single channel that carries Ethernet signals to all stations. No matter how the media segments are physically connected together, only one signal channel is delivering frames over those segments to all stations on a given Ethernet.

1.5.1   Ethernet Bus Topology

The *bus topology* was the original layout used for Ethernet. Early bus topologies consisted of a long central coax cable with workstations connected in sequence using T-connectors. The lower interface on the *T* provided a link to a single workstation, whereas both the left and right interfaces allowed for continuation of the bus cable to the next T-connector (workstation) in line. A special bayonet socket was typically used to terminate the bus at both the head and tail end of the coaxial cable. This topology is not used any more because it is bulky and inefficient.

The following are advantages of the Ethernet bus topology:

- Linear layout is inexpensive and easy to implement.

- Less cabling is required when compared to other topologies.

- New nodes (PC workstations) may be added without disrupting network traffic.

The Ethernet bus topology has several disadvantages as well, including the following:

- Troubleshooting is difficult due to an unstructured wiring scheme that provides no central point of cable concentration.

- The lengths of the cable segments between T-connectors need to be a certain length or multiples of a certain length.

- A break in the bus may be fatal to the LAN or create two isolated segments of a LAN.

See Figure 1-4 for a depiction of an Ethernet bus topology.

1.5.2   Ethernet Star Topology

The *star topology* is becoming the most common Ethernet topology in use today because it lends itself to a central distribution point for cabling and traffic aggregation. Often referred to as a *wiring closet*, this central connec-

Figure 1-4
Ethernet bus
topology

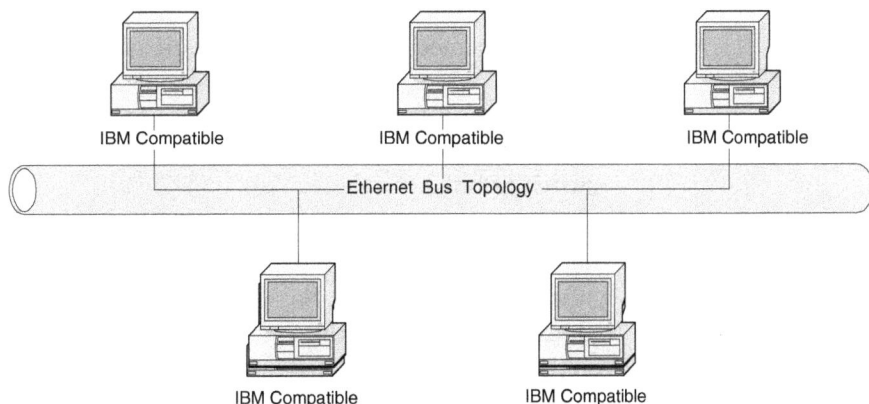

IBM Compatible   IBM Compatible   IBM Compatible

——————————————Ethernet Bus Topology——————————————

IBM Compatible   IBM Compatible

tion is where the data cables from offices and cubes terminate into a modular RJ-45 patch panel for easy moves, adds, and changes. Typically, links are wired out from the closet (or closets) through the walls, floors, or ceilings in a star pattern. Then RJ-45 patch cables connect the cable terminations from the patch panel to the LAN hub/switch in the wiring closet (refer to Figure 1-1). If one arm of the star breaks (one workstation connection becomes defective), it does not affect any other stations on the LAN. Floor-to-floor backbone connections typically run at higher speeds and aggregate the entire enterprise's LAN traffic into a single, central data communications or server room. As previously discussed, this topology is the collapsed backbone (refer to Figure 1-3).

The following are advantages of the Ethernet star topology:

- A simple design that requires minimal preplanning.
- Single connection failures to workstations do not affect the rest of the network.
- The switch/hub is a central point for isolating network problems. Most Ethernet hubs and switches in wire closets have *light-emitting diode* (LED) indicators that will indicate where a link failure has occurred. The link failure could be caused by any one of several problems:
  - A failed NIC in a workstation
  - An unplugged workstation cable
  - A workstation cable that has become defective
  - A defective port on the switch (not all outages are workstation related)

The Ethernet star topology has several disadvantages as well, including the following:

- If a wire closet switch/hub fails, the network goes down for all users who are connected to that switch/hub. A large building such as a high-rise MTU will usually have at least one switch or hub in a wiring closet on each floor. In this scenario, a switch or hub failure would impact all the users on that floor who are connected to that particular switch or hub.
- Implementation requires large amounts of cable. However, it should be noted that about 75 percent of the cost to pull cables to cubes and offices is labor charges. So when expanding a new office or renovating an existing one, plan the network design in a forward-looking manner so that all required cables can be pulled to each workstation just once. The general rule of thumb is to always pull more cable to each

Figure 1-5
Star logical network
topology (physical
cabling layout not
depicted)

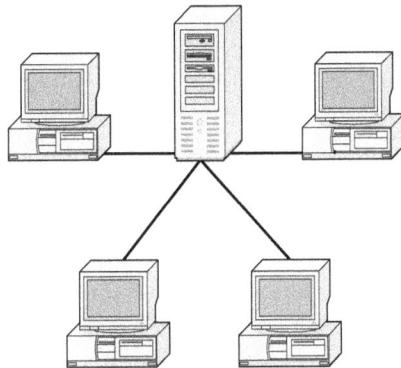

workstation than you think you'll need, provided you have enough
budget and space to accommodate the extra cable(s). It will cost
considerably more money to pull additional cable at a later date. See
Figure 1-5 for a depiction of a star Ethernet topology.

1.5.3   Ethernet Tree Topology

The *tree topology* is essentially a hybrid of the bus and star layouts. The
basic topology is similar to that of a bus, with nodes connected in sequence
to a linear central cable. Tree networks usually have branches that contain
multiple workstations that are connected point to point in a star-like pattern. Signals from a transmitting node travel the length of the medium and
are received by all other nodes.

1.6   Carrier Sense Multiple Access and Collision Detection (CSMA/CD)

Ethernet uses the CSMA/CD protocol for media (transmission) access.
CSMA/CD is specified in the IEEE 802.3 standard.

1.6.1   Carrier Sensing

In an Ethernet system, any device can try to send an Ethernet frame at
any time. Each device senses whether the line (signal channel) is idle and

therefore available to be used. If the line is idle, the device begins to transmit its first frame. If another device has tried to transmit at the exact same time, an electronic collision is said to occur and all frames from both nodes are discarded. Each device then waits a random amount of time and retries until it's successful in sending its transmission. To reduce the probability of collisions, an Ethernet interface—a workstation or server—cannot transmit when another one is transmitting. Since signal propagation takes time (signals do not travel faster than the speed of light), some time elapses between the beginning of the transmission and the time a collision might occur from an interface, which had not yet begun receiving the transmitted signal. Both interfaces believe they were the first to begin transmitting and they are both right in their own frame of reference. This period of uncertainty is known as the *contention time* and is twice the propagation time between the most distant pair of interfaces in the Ethernet network. This is why Ethernet technology is known as a *contention-based system*. Propagation time is about 5 nanoseconds per meter on a coaxial cable and greater when using copper UTP.

Key: A system employing CSMA/CD will thus be in one of three states: transmission, contention, or idle.

A small amount of idle time is required before and after each frame transmission and contention state for the carrier sensing to do its job. However, in a busy network, several interfaces may attempt to transmit when they sense that the network is idle.

1.6.2   Collision Detection

Suppose an Ethernet-enabled device (such as a workstation) broadcasts frames that are all the same size (for simplicity sake). The amount of time to transmit a frame, the frame time, is the ratio of the frame length and bit rate. A collision will result with a given frame if any other frame is broadcast within one frame time of the start of the given frame. This window of vulnerability is two frame times in duration since the collision may be with a frame that started before the given frame or after it. To recover from the collision, the network interface waits a short while and then retransmits. If a collision occurs on the second attempt, it waits twice as long as it did the first time before retransmitting. This process is called *backoff*. The retransmission is attempted about six times, and then the data to be transmitted is simply discarded. Hence, a busy Ethernet network will drop many packets.

In a real data network, things are not as tidy as in the statistical studies. Frames vary in size, and network traffic tends to come in bursts rather than being fairly and uniformly spread out. Contention comes only at the beginning of the frame so the simplest measure of the network load as it relates to collisions is the packet rate (or data rate).

In the extreme case, an Ethernet network can theoretically achieve a condition called *collapse* in which it does not leave the contention state for any significant amount of time. In theory, it can be triggered by a bad combination of timing with just a handful of interfaces all initiating a transmission as soon as the network appears idle, but the probability is astronomically low that such an event would occur. Typically, collapse is triggered by an NIC that has a failure in its circuitry. Sometimes a single NIC will jabber by sending a continuous stream of bad frames (often without stopping to do carrier sensing or collision detection). In industry parlance, this is also sometimes known as a *broadcast storm* caused by a *screaming NIC*.

Key: The greater the number of nodes in an Ethernet network, the greater the impact of collision detection on the network.

1.6.3 Throughput

CSMA/CD is not always predictable. Although it works well for a small number of nodes per wire (or a large number of quiet nodes), it tends to lose throughput if the network gets too busy. In data transmission, throughput is the amount of data moved successfully from one place to another in a given period of time.

On a busy CSMA/CD network, throughput could be limited to less than 40 percent of the available bandwidth. So in a 10 Mbps network, average throughput could be only 4 Mbps. Above that, packets spend all their time colliding and backing off and not getting any real transmitting done! Eventually, no complete packets can go through, and the throughput actually drops to zero. One way to get around this shortcoming of Ethernet is to dedicate a switch port to each device. This effectively makes CSMA/CD a nonissue since no packets are present to collide with on a private cable. The LAN Ethernet switch interleaves all frames from all nodes together and then routes the traffic upstream to the central data communications or computer room.

Key: Another throughput solution is to reduce delays caused by contention for the wire by separating the nodes into different collision domains. For example, *virtual LAN* (VLAN) broadcast domains could be used. VLANs will be reviewed in the section "Virtual LANs (VLANs)."

In summary, as more hosts place demands on the network, the share of the total available bandwidth that each host gets decreases. In addition, as the demand for network bandwidth increases, network efficiency begins to drop. Figure 1-6 depicts a CSMA/CD LAN.

1.7   Ethernet Hardware, Hubbing, and Switching

This section reviews the basic hardware elements required when constructing and operating an Ethernet LAN. It also includes a review of the fundamentals of Ethernet switching and VLANs.

Figure 1-6
CSMA/CD LAN

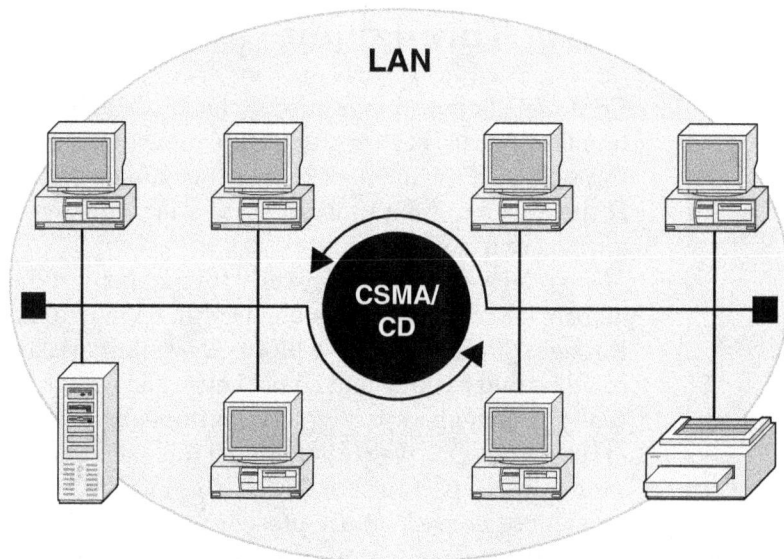

1.7.1 Network Interface Cards (NICs)

As discussed earlier, installing an NIC enables a computer workstation to communicate with the network. Ethernet NICs are specially designed for each particular network type and/or protocol (for example, a 10BaseT NIC or 100BaseF NIC). Many different speeds and intelligence choices are available, so the decision on which one to buy is part of the total network design.

Key: There is an increasing trend toward including an Ethernet NIC in most new desktop computers sold by consumer-focused electronics and computer retailers. This underscores the fact that Ethernet has wide appeal, and its use in home networking will only continue to grow. Home networking (via Ethernet) is expected to grow at a *compound annual growth rate* (CAGR) of 54 percent from 2001 to 2006.

1.7.2 Ethernet LAN Hubbing

Ethernet was designed to be easily expandable to meet the networking needs of a given site. To help extend Ethernet systems, equipment vendors sell Ethernet switches that provide multiple Ethernet ports and that can easily scale. One option is to cascade Ethernet switches together in a wiring closet where they're linked together in a daisy-chain format, with one of the switches being designated as the true hub that back hauls all traffic terminating into that particular wire closet onto the LAN backbone. Once on the backbone, the traffic moves upstream to the main computer room.

Two major kinds of hub are available: repeater hubs and switching hubs. Each port of a repeater hub links individual Ethernet media segments together to create a larger network that operates as a single Ethernet LAN backbone. The total set of segments and repeaters in the Ethernet LAN must meet the roundtrip timing specifications of Ethernet. The second kind of hub provides packet switching, typically based on bridging.

Key: Unlike a repeater hub whose individual ports combine segments together to create a single large LAN, a switching hub makes it possible to divide a set of Ethernet media systems into multiple LANs (also known in this context as *collision domains*) that are linked together by way of the packet switching electronics in the hub. These separate collision domains are VLANs. Although an individual Ethernet LAN may typically support anywhere from several dozen or several hundred computers, the total system of Ethernet LANs linked with packet switches at a given site may support many hundreds or thousands of machines.

1.7.3 Repeaters

Multiple Ethernet segments can be linked together to form a larger Ethernet LAN using a signal amplifying and retiming device called a *repeater*. Through the use of repeaters, an Ethernet system of multiple segments can grow as a nonrooted branching tree. This means that each media LAN segment is an individual branch of the complete signal system. Even though the media segments may be physically connected in a star pattern with multiple segments attached to a repeater, the logical topology is still that of a single Ethernet channel that carries signals to all stations. *Nonrooted* means that the resulting system of linked segments may grow in any direction and does not have a specific root segment. Most importantly, segments must never be connected in a loop. Every segment in the system must have two ends since the Ethernet system will not operate correctly in the presence of loop paths.

Key: In modern GigE metro networks, repeaters may be used in the middle of point-to-point circuits to extend the distance of these circuits up to approximately 60 to 70 miles end to end. Total permissible distance in these circuits is dependent on the age of the optical fiber and the total number of splice points end to end.

1.7.4 Ethernet LAN Switching

Early Ethernet designs and operations were initially limited to many users sharing a single 10 Mbps bandwidth pipe. This occurred by having the connections from all workstations on a LAN terminate into an Ethernet hub in the wiring closet on each floor of a high-rise building (or in dispersed wiring closets in a large, one-story building). This hub operated on a shared basis, meaning that Ethernet traffic from all workstations terminated into the hub, and the traffic from all the workstations had to share the 10 Mbps pipe for transport onto the LAN backbone. The traffic was then routed over the LAN backbone to the main data or communications room. This setup led to decreased throughput, but as applications began to demand more bandwidth, equipment designers started to provide new solutions. By providing switched Ethernet, each device could have a dedicated 10 Mbps connection into the switch. The switch then interleaved the traffic from all the dedicated ports together into one massive data stream, again, back to the central computer room. Switching technology was adopted very rapidly due to its obvious transmission benefits to the LAN environment. Performance improves in LANs where LAN switches are installed because the LAN switch isolates collision domains. By spreading users over multiple collision domains, collisions are decreased and performance improves. Many LAN switch installations assign just one user per port, which gives that particular user an effective dedicated bandwidth of 10 (or 100) Mbps. Switches also allowed for the development of VLANs. As these switches needed to be connected to the LAN backbone, the backbone would have to run at much higher speeds than 10 Mbps because of the huge amount of traffic they aggregated. This led to the development of Fast Ethernet, a 100 Mbps version of Ethernet.

1.7.4.1 Understanding Switching Basics In the data communications world, the term *switching* was originally used to describe packet-switched technologies, such as frame relay, *Switched Multimegabit Data Service* (SMDS), and X.25. Today, switching refers to a technology that is similar to a bridge in many ways.

The term *bridging* refers to a network technology in which a device (a bridge) connects two or more LAN segments together. A bridge transmits datagrams from one network segment to their destinations on other network segments. When a bridge is powered and begins to operate, it examines the MAC address of the datagrams that flow through its switching matrix in order to build a table of known destinations. If the bridge knows that the destination of a datagram is on the same segment as the source of the datagram, it drops the datagram because there is no need to transmit (forward) it. If the bridge knows that the destination is on another segment, it retransmits the datagram on that segment only. If the bridge does not know the destination segment, it transmits the datagram on all segments except the source segment. This technique is known as *flooding*, which is also known as a *broadcast transmission*.

Like bridges, switches connect logical LAN segments, use a table of MAC addresses to determine the segment on which a datagram needs to be transmitted, and reduce overall network traffic.

1.7.4.2 Switching in an Ethernet Environment An Ethernet LAN switch improves throughput by separating collision domains and selectively forwarding traffic to the appropriate segments. Figure 1-7 shows the topology of a typical Ethernet network in which a LAN switch has been installed.

Figure 1-7 shows that each Ethernet segment is connected to a port on the LAN switch. If Server A on Port 1 needs to transmit to Client B on Port 2, the LAN switch forwards Ethernet frames from Port 1 to Port 2, thus sparing Ports 3 and 4 from frames destined for Client B. If Server C needs to send data to Client D at the same time that Server A sends data to Client B, it can do so because the LAN switch can forward frames from Port 3 to Port 4 at the same time it is forwarding frames from Port 1 to Port 2. If Server A needs to send data to Client E, which also resides on Port 1, the LAN switch does not need to forward any frames. You get the idea!

Figure 1-7
Ethernet switching
operations

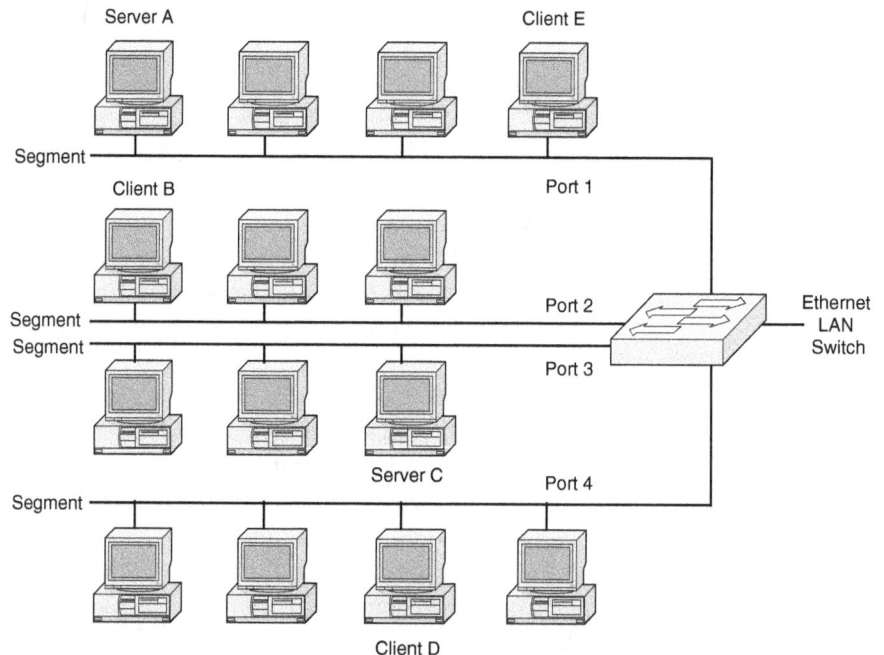

Key: Switches operate at much higher speeds than bridges and can support functionality known as VLANs.

1.7.5 Virtual LANs (VLANs)

VLANs are formed to group related users together, regardless of the physical connections of their hosts (that is, workstations and/or servers) to the network. The users can be spread across a campus network or even across geographically dispersed locations. A variety of strategies can be used to group users together. For example, the users might be grouped according to their department or functional team. In general, the objective is to group users into VLANs so that most of their traffic stays within one VLAN—*their* VLAN. In other words, the idea is to group users where most of the traffic is confined, or clustered, among the users themselves. VLANs offer the following benefits to a network:

- **Broadcast control**   Just as switches physically isolate collision domains for attached hosts and only forward traffic out a particular port, VLANs provide logical collision domains that confine broadcast and multicast traffic to the bridging domain.

- **Security**   If a router is not included in a VLAN topology, no users of that VLAN can communicate with the users in another VLAN, and vice versa. This extreme level of security can be highly desirable for certain projects and applications.

- **Performance**   Users who require high-performance networking can be assigned to their own VLANs. For example, an engineer who is testing a multicast application and the servers the engineer uses could be assigned to a single VLAN. The engineer experiences improved network performance by being on a dedicated LAN, and the rest of the engineering group experiences improved network performance because the traffic generated by the network-intensive application the engineer used is isolated to another VLAN.

- **Network management**   Software on the Ethernet LAN switch enables you to assign users to VLANs and, at a point later in time, reassign them to another VLAN. Recabling to change network connections is no longer necessary in the switched LAN environment because network management tools allow for logical reconfiguration of the LAN in seconds. Physical reconfiguration (via cabling) is no longer necessary.

1.8   Ethernet Versus Token Ring

During the late 1980s and early 1990s, a battle occurred within the information technology industry regarding what LAN operating system was superior: Ethernet or another LAN operating system known as Token Ring. In this context, *superior* translates to cheaper and more manageable. As this section (and this book) illustrates, the Ethernet standard has won this battle hands down.

1.8.1 Token Ring

Token Ring (IEEE 802.5) is a token-passing protocol used in the LAN environment. Originally specified to transmit data across copper cable at a rate of 4 Mbps, updated versions of Token Ring transmit at a rate of 16 Mbps.

Workstations or servers connect to a Token Ring network using individual lobe cables that attach to a *multistation access unit* (MSAU). The MSAU acts as the hub of the Token Ring network, and patch cables connect MSAUs together to form the *logical* ring.

Although Ethernet overwhelmingly remains the most popular LAN technology, Token Ring ranks second in local network implementations. Token Ring networks make use of a token-passing process that was initially developed by IBM to address the collision problems found in Ethernet networks. In token-passing networks, access to the network is controlled by the use of a special electronic packet, or token, which is passed in one direction from node to node around a closed, logically circular network layout. The data rate for Token Ring can be either 4 or 16 Mbps, but not a mixture between the two. Figure 1-8 shows a depiction of the logical transmission flow in a Token Ring LAN.

In order for a node to transmit in a token-passing network, it must seize the token, alter one bit to mark it as busy, insert the payload information it wants to transmit, and send it to the next node on the ring. The information packet circulates the ring node to node until it reaches the destination node. The destination node copies the information and sends the token back onto the ring toward the sending node. The sending node checks to see if the information made it to the destination without errors. If no errors are found, a new token is released onto the ring. Generally, a node obtains a token from an upstream node and passes it to a downstream node. Each node may only possess the token for a specified period of time before returning it to the network for other nodes to use.

The shared nature of token passing makes the Token Ring protocol deterministic compared to Ethernet, which is contention based. This means

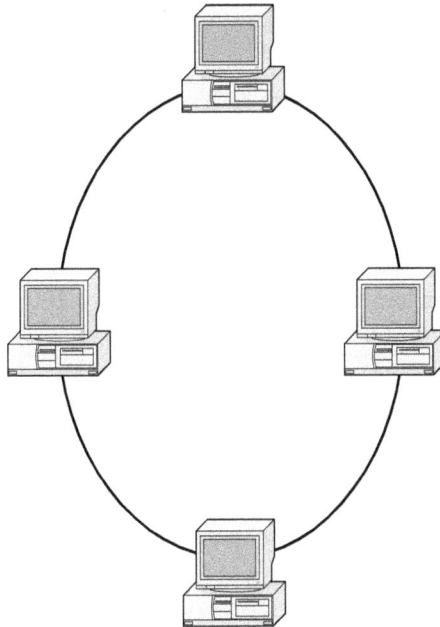

Figure 1-8
A Token Ring LAN
(logical transmission)

that Token Ring LANs enable network managers to calculate and predict the maximum time interval (rotational delay) between node transmissions. As a result, token-passing networks are ideally suited for mission-critical applications or any application that requires predictable delay before timing out without a consistent service response time.

The following are advantages of token-passing networks in comparison to Ethernet:

- Capability to prioritize transmissions and built-in diagnostics.
- Use of a shared token eliminates network contention and supports better performance during periods of high traffic.
- The large packet length (on a 16 Mbps ring) allows for more efficient use of bandwidth.
- Flexible media choices (UTP, coaxial cable, and optical fiber) allow for easy design and implementation.
- Token-passing networks offer great flexibility when it comes to transmission media. Distance limitations of the media are also generous. For example, a single Token Ring may theoretically be 120 miles in length.

The following are advantages of Ethernet networks in comparison to token-passing networks:

- Ethernet has already proven to be scalable, with high performance.
  - Technology migration has moved from 10 to 100 to 1,000 Mbps (GigE) and now 10,000 Mbps (10 GigE). 100 GigE is not unrealistic at some point in the future.
  - It is full-duplex, symmetrical, and switched (versus shared).
  - Ethernet has longer-range distances, up to 150 km ($\approx$90 miles) when used in fiber-based LX mode.
- Ethernet has become the de facto LAN standard because it's less expensive to implement and operate. Ethernet is also poised to become very popular in MAN environments.
  - Ethernet benefits from 26 years of evolution.
  - Ethernet dominates over 90 percent of installed networks worldwide.
  - There is a large pool of trained network professionals who understand Ethernet.
  - There is wide adoption of Ethernet standards across the world and through various standards bodies. Hundreds of millions of desktops across the world are equipped with Ethernet NICs.

1.9   Drivers of Ethernet Growth

Fast Ethernet was standardized in 1995, providing a complete solution for the campus network. It also allowed for improvements in LANs that operated in MTUs. The introduction of Fast Ethernet was a watershed event for the LAN community and network managers everywhere. Ethernet technology was now a scalable technology, both in speed and distance. Network managers could mix or match speeds of 10 and 100 Mbps within the appropriate parts of the network to optimize its function.

Today's LANs are becoming increasingly congested and overburdened. In addition to a constantly growing population of network users and applications that require increasing amounts of bandwidth, several factors have combined to stress the capabilities of traditional LANs:

- **Faster processors (*central processing units* [CPUs])**   In the mid-1980s, the most common desktop workstation was the PC. At the time, most PCs could execute 1 *million instructions per second* (MIPS). Today, workstations with 50 to 75 MIPS of processing power are common, and *input/output* (I/O) speeds have increased accordingly. Two modern engineering workstations (CAEs) on the same LAN can easily saturate the network.

- **Faster operating systems**   Until the late 1990s, operating system design had constrained network access. Of the three most common desktop operating systems (Windows, UNIX, and Mac OS), only the UNIX operating system could multitask. Multitasking enables users to initiate simultaneous network transactions. The release of Windows 95 in 1995 reflected a redesign of Windows that included multitasking, and PC users could then increase their demand for network resources.

- **Network-intensive applications**   Use of client-server applications, especially the Internet, is increasing. Client-server applications enable network administrators to centralize data, which makes it easy to maintain and protect. Client-server applications also free users from the burden of maintaining information and the cost of providing enough disk space to store it. Given the cost benefit of client-server applications, these applications are likely to become even more widely used in the future.

Switching is a technology that alleviates congestion in Ethernet and Token Ring LANs by ultimately reducing traffic and increasing bandwidth. LAN switches are designed to work with existing cable infrastructures so that they can be installed with minimal disruption of existing networks. Installation of LAN Ethernet switches usually replaces shared Ethernet hubs.

In summary, Ethernet has matured into the dominant LAN technology of today. According to IDC, the Ethernet customer base adds about 30,000,000 users per year. All popular operating systems and applications are Ethernet compatible, as are the upper-layer protocol stacks such as TCP/IP, *Internetwork Packet Exchange* (IPX), DECnet, and so on. The main reasons for the dominance of Ethernet are

- High degree of reliability

- Availability of management tools

- Scalability—the ability to increase data rates easily in small or large increments

- Lower cost

- Ease of deployment (autosensing capability)

Now that it's apparent that Ethernet has become the dominant LAN standard in the world's LANs, let's explore how and why it's making its venture into the MAN.

Metro Area
Ethernet

2.1   Metro Area Networks (MANs)

A *metropolitan area network* (MAN) is a packet-based, wide area data network that is specifically designed for interconnecting separated *local area networks* (LANs) across a limited geographic area. The area may be as small as the financial center of a large city or as large as several cities and the suburbs around them, but it is not a service designed to be continental or international in reach. The MAN *core* is the larger intracity network connecting *central offices* (COs), carrier hotels, and other *points of presence* (POPs) within an urban area. The access part of the MAN consists of the *last-feet links* connecting the customer premises to the core. Don't misunderstand the term *last feet*—these last feet could be hundreds or even thousands of feet long.

Network service providers face a variety of challenges as they seek to capitalize on the opportunities resulting from emerging technologies, advances in network standards, and more demanding user requirements. Telecommunications industry deregulation has resulted in increased competition, has certainly stimulated innovation, and has helped to reduce service prices. At the MAN level, there is now tremendous pressure for expanded capacity to support broadband local access and high-speed *wide area networks* (WANs), especially the Internet. All of these factors suggest that a flexible, proven MAN architecture combined with multivendor compatibility is urgently needed so that new services can be introduced by local carriers.

Metropolitan networks are undergoing a radical transformation. Combined with the demands of new applications, developments in last mile and fiber-optic technologies have put pressure on service providers and infrastructure vendors to improve their services and products. Figure 2-1 illustrates the basic elements of a metro network and its relationship to access and long-distance networks.

Since the late 1990s, the MAN has emerged as a critical and dynamically evolving arena within the overall network infrastructure. Not only are MAN traffic demands rapidly escalating, but the underlying network architectures, protocols, and technologies are also experiencing sweeping change.

The emergence of *wavelength division multiplexing* (WDM), the rise of higher-speed optical connections, and the drive toward voice and data convergence are all combining to put pressure on existing network architectures to keep up with the parallel explosions in both demand and capacity. In addition, the competitive landscape for MAN service providers is shifting

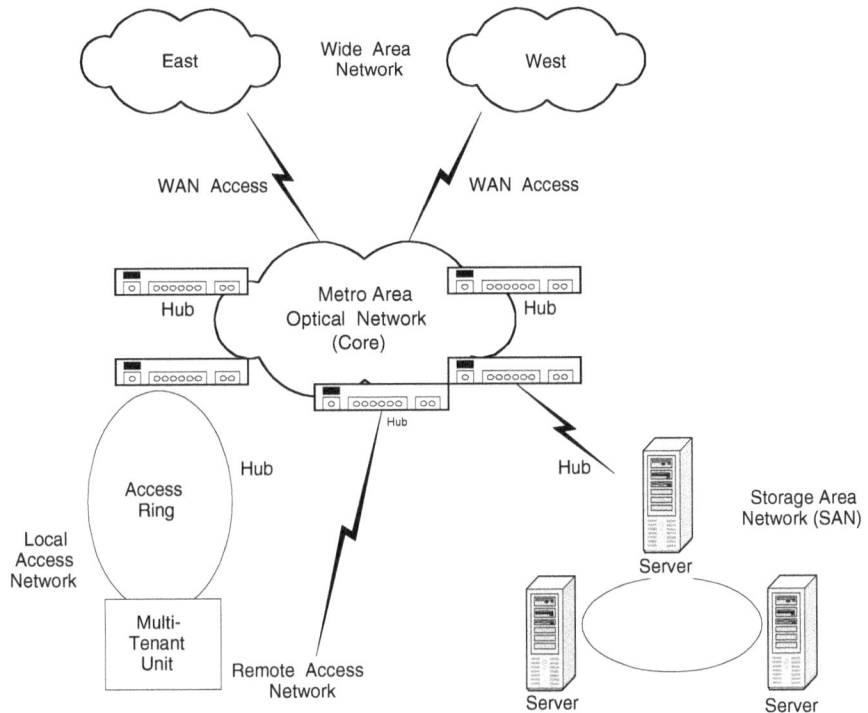

Figure 2-1
LAN-MAN-WAN
network integration

with the influx of whole new classes of carriers who do not carry the legacy baggage and inertia of previously deployed infrastructures, such as *Synchronous Optical Network / Synchronous Digital Hierarchy* (SONET/SDH).

Customer demand for broadband voice and data services has exploded. A new world of bandwidth-hungry, multiprotocol services is providing a tremendous opportunity for both incumbent and emerging carriers. Services such as videoconferencing, *Internet Protocol* (IP) telephony, storage area networking, Internet audio, and telecommuting—to name a few—may replace standard voice and dial-up 56 Kb data links as the dominant profit centers by 2007.

Although local services such as Internet access consume increasing amounts of bandwidth, metro and long-haul core networks are rapidly developing the potential for dramatic improvements in bandwidth efficiency. With the advent of technologies such as WDM and specifically *dense wavelength division multiplexing* (DWDM), today's fiber technology can deliver in excess of 1 *petabit per second* (Pbps) over one single optical fiber. A petabit is 1 quadrillion bits, or the equivalent of 1,000 *terabits* (Tb). A new generation of services we call *metro access ramps* promises to relieve this congestion and offer carriers ways to create competitive advantage in the complex, high-stakes game of provisioning converged voice and data services.

Key: The developing convergence of high-bandwidth applications originating from the user and the existence of high-bandwidth capacity in the long-haul core (interstate and intercontinental networks) has placed a tremendous amount of pressure on the edge of metro networks. Until recently, MAN infrastructure has not kept pace with improvements in LAN and WAN network infrastructures. As both the LAN and WAN made significant improvements in bandwidth capabilities, the MAN remained stagnant. MAN connections were mainly restricted to DS-1 or DS-3 private lines, using *time division multiplexing* (TDM) technology. For many users, a T1 didn't provide enough bandwidth, but a DS-3 was too expensive. Because of these service-offering limitations, a network bottleneck has now emerged in metro access (and core) networks. A new breed of optical metro access and transport technologies, such as *Gigabit Ethernet* (GigE), promises to break this bottleneck between end users and the long-haul "core." Between 2002 and 2007, the architecture of the metro public network will have to change dramatically, if gradually, to *remove* this latest bottleneck in the end-to-end network.

Figure 2-2 illustrates the congestion problem in today's MAN.

2.1.1   Explosive Demand Growth

Bandwidth growth has also been explosive in the LAN, propelled by the availability and deployment of GigE since 1998. WAN network capacity has also exploded, fueled by the 300 percent annual growth in Internet traffic. The remaining network link has been between the LAN and WAN: the MAN.

Figure 2-2
The MAN is the "choke point" in today's PSTN.

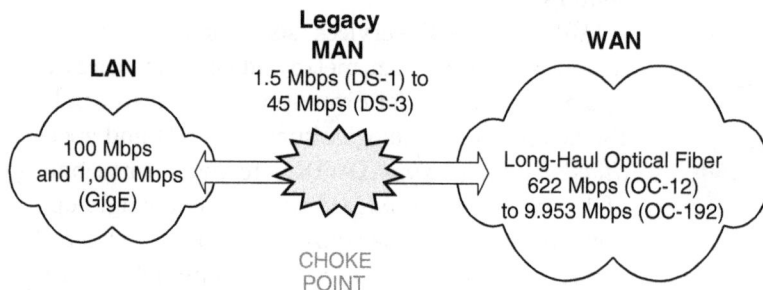

LAN
100 Mbps and 1,000 Mbps (GigE)

Legacy MAN
1.5 Mbps (DS-1) to 45 Mbps (DS-3)

CHOKE POINT

WAN
Long-Haul Optical Fiber
622 Mbps (OC-12) to 9.953 Mbps (OC-192)

In addition to the millions of new users who are going online every month, the nature of Internet applications are becoming more bandwidth intensive as the Internet has become an increasingly visual environment and multimedia content has become more widespread. For users, the emergence of broadband connectivity in the form of cable modems and *Digital Subscriber Line* (DSL) has also significantly expanded the capacity of the last mile of the network pipeline, giving them faster access and greater inclination to make use of bandwidth-intensive content on the Internet.

To underscore the dramatic evolution of today's MAN, consider the following statistics and projections, but bear in mind that the information provided in the following list may not be consistent due to disparate research methods and sources. In any case, even though there may be some disparity in the following stats and facts, one thing is certainly consistent: All researchers and analysts are showing tremendous growth for metro area (Gigabit) Ethernet services well into the near future.

- Industry analysts and recent market research depict a dramatic increase in end-user metro data consumption into the year 2006. A recent report issued by Infonetics Research shows a 552 percent growth rate for metro data products, and estimates that (enterprise) customers in the United States and Canada will increase spending from $420 million to $2.7 billion between 2002 and 2006.

- According to Forrester Research, business demand for Internet access bandwidth in the United States will grow from an estimated 296 Gbps in 1999 to an estimated 3,640 Gbps (3.6 Tbps) in 2003—a *compound annual growth rate* (CAGR) of 87 percent!

- The growth of data services is at more than 50 percent a year and the growth of voice is at more than 15 percent a year.

- BancBoston's Robertson Stephens predicts that to support the growth of the Internet alone, 350,000 additional DS-1s and 25,000 additional DS-3s must be provisioned by 2005, compared to the current installed base of approximately 300,000 DS-1s and 2,200 DS-3s.

- Driven by web hosting, and *business-to-business* (B2B) and *business-to-consumer* (B2C) e-commerce, IDC projects that by 2003, 55 percent of all enterprises and 71 percent of all home businesses will be online. This is predicted to result in an IP services market of $5.75 billion and a line aggregation market (such as *Asynchronous Transfer Mode* [ATM], frame relay, optical Ethernet, and private line) of $21.5 billion by 2003.

- As corporations deploy Internet *virtual private networks* (VPNs) to replace existing private-line and frame relay networks, the U.S.

Internet access market is expected to grow from less than $3 billion in 1998 to $42 billion by 2003.

- Backbone carriers, the largest being AT&T, WorldCom, Sprint, Level 3, and Qwest, terminate their facilities at various hubs within any particular metro. Interconnection must be supported across different backbone networks and through various local access channels. As a result, connecting carrier hubs within a metro generates a very large area of demand that looks to grow from 2,000 OC-48 equivalents in mid-2001 to 34,000 OC-48 equivalents in 2005, representing a 75 percent CAGR.

- Enterprise traffic is currently very concentrated, as in a typical Tier 1 *Metropolitan Statistical Area* (MSA). For instance, 300 *multitenant units* (MTUs) out of more than 15,000 constitute 80 percent of data revenues today.

- Translating bits into dollars, metro data traffic generated approximately $15 billion of revenue in 2000. By 2005, data revenues are forecasted to increase by $35 billion, creating a $50 billion market, a 35 percent CAGR that factors in an anticipated aggregate 30 percent annual unit price decline (in cost per port) versus bandwidth growth.

- It's projected that from 2001 to 2005, the market for GigE ports sold into the carrier market will grow from $350 million in 2001 to over $4.4 billion. The market will experience its highest growth rates in 2002 and 2003, by which time the market for 10 GigE in both the LAN and WAN will be in full swing. The forecast reflects a 30 to 35 percent a year decline in the cost of Ethernet ports offset by a healthy growth rate in the number of ports sold.

- According to JP Morgan, the local data market generated revenue of $17.6 billion in 2000 and is projected to grow 22 percent annually to $48 billion by 2005.

- Another study says that aggregate voice, data, and Internet traffic demand in metro areas is projected to grow to $50 billion in 2005. Demand for bandwidth in the metro will continue to grow rapidly, fueled by transport needs of data centers, carrier interconnection, and large enterprises. Much of this bandwidth demand will depend on adoption rates for Internet applications (such as *storage area networks* [SANs], application hosting, and so on), technology cost trends, and the tendency to purchase enough new connectivity to maintain *quality of service* (QoS) as demand grows. Traffic is concentrated, with the top 15 metros accounting for almost 80 percent of total demand and a select few customers generating the majority of traffic flow.

- A Lehman Brothers study released in August 2001 predicts that the Ethernet protocol, notably Fast Ethernet and GigE, should become the principal access protocol for enterprises by 2006. It's suggested that Ethernet should account for 60 percent of total bandwidth utilization, due to its low-cost structure and familiarity in IT enterprise *netvironments*.

- In-Stat/MDR expects the metro optical network market to grow from $13 billion in 2001 to $23.6 billion in 2005, making it one of the fastest growing segments in the telecommunications industry.

- IDC also predicts that the metro Ethernet market will experience a shift toward higher port speeds by 2006. In 2001, 100 Mb and 1 GigE ports account for approximately 90 percent of metro Ethernet revenue. IDC believes this percentage will drop to 78 percent in 2006 due to the introduction and deployment of 10 GigE port speeds around 2003.

- The overall amount of metro circuits (ports) is growing at approximately 65 percent, fueled by the growth of data centers and carrier hotels.

- Next-generation SONET technologies (such as SONET-lite, metro DWDM, and *Ethernet over SONET* [EoS]) are 30 to 70 percent more cost efficient than legacy networks (such as TDM and ATM).

- Data center traffic is growing at nearly 100 percent annually and will consume 40 percent of the total metro bandwidth by 2005. These data centers happen to be concentrated in the top 15 Tier 1 metros—the same metro areas that account for 80 percent of traffic demand. In fact, the top 4 markets constitute approximately 40 percent of all MAN traffic.

- The top 15 metropolitan areas in the United States account for 80 percent of total bandwidth demand. A select few customers (such as *Internet service providers* [ISPs] and Fortune 100 companies) are generating the majority of this traffic flow.

- The CAGR for the MAN marketplace is predicted to be 26 percent through 2006, according to IDC. Table 2-1 illustrates the growth rate year by year.

- In 2000, the total demand for metro bandwidth was satisfied by the equivalent of approximately 8,000 OC-48 circuits. To put this into perspective, that's the equivalent of 258 million DS-0 circuits. By 2005, metro traffic will likely require more than 100,000 OC-48 equivalents, with service providers, data centers, and enterprises driving the majority of the growth. Using the DS-0 perspective, that's the

Table 2-1

Ethernet ports
forecast—2001
to 2006

U.S. Metropolitan Ethernet Service Ports and Customers—2001 to 2006							
Year	**2001**	**2002**	**2003**	**2004**	**2005**	**2006**	**CAGR**
Ports	5,148	7,155	9,373	12,091	15,295	19,119	**30%**
Customers	2,629	3,575	4,648	5,810	6,972	8,367	**26%**
Ports per Customer	2.0	2.0	2.0	2.1	2.2	2.3	**3.1**

Key assumptions:
- The data include *transparent LAN* (TLAN), Ethernet-based MAN transport, and Ethernet-based Internet access.

Messages in the data:
- Ethernet will displace some DSL and *Integrated Service Digital Network* (ISDN) implementations at lower speeds, and T-1 and T-3 at higher speeds.
- Because Ethernet is currently primarily a point-to-point technology, IDS expects it to have minimal effect on Frame Relay and ATM in the near term.

Source: IDC (2001)

equivalent of 3.2 billion DS-0 circuits! That represents an increase of over 1,200 percent in a five-year period.

Figure 2-3 shows a prediction for Ethernet access revenue in the United States into 2006.

Ethernet will grow rapidly as an access service, but probably not to billions of dollars in service revenue by 2005 as some analysts predict. The biggest impediment to this growth is the expenditure needed to connect new buildings to the network. This is also known as installing fiber *laterals* off of existing metro core rings. This issue will be fully reviewed in Chapter 3. As data demands grow and new laterals are ultimately built, Ethernet will increasingly be used to provide high-speed Internet connections. However, with copper cable already going into every building, the T1 is not about to lose its grip on the access network. T1 circuits will remain popular with smaller enterprises.

Key: The evolution of the metro access and transport landscape will prove challenging for new entrants hoping to exploit GigE technologies. Success will be predicated on flawless execution, high building penetration rates, and the degree to which *Incumbent Local Exchange Carrier* (ILEC) efforts in this market can be overcome.

Figure 2-3
U.S. Ethernet access
service revenue
(Source: America's
Network)

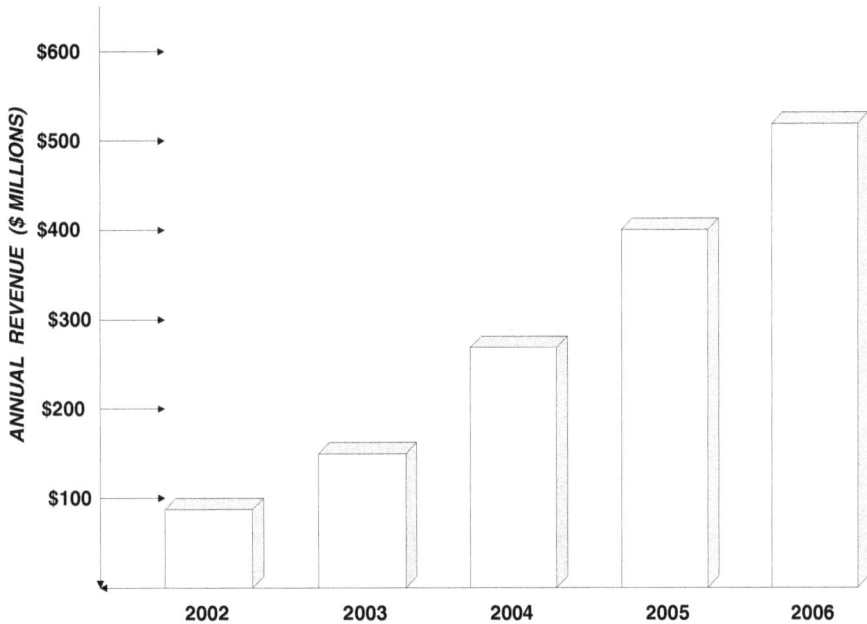

Figure 2-3
U.S. Ethernet access service revenue (Source: America's Network)

2.1.2 Services and Technology Consolidation

Service convergence is another key factor driving the ongoing evolution of MANs. Historically, metropolitan communications infrastructures have been created and optimized for voice traffic, while data requirements have arisen as something of an afterthought. But since 1995 (the "Year of the Internet"), Internet growth has begun to skyrocket and data network infrastructures have become more important with every passing year. From the copper-loop connections between users and their COs to the uniform TDM infrastructures across the metro and wide area, all aspects of the traditional *Public Switched Telephone Network* (PSTN) were designed for *connection-oriented* voice traffic rather than *connectionless* data traffic. As data has become the dominant traffic type—and it has—a driving need has arisen for the seamless convergence of data, toll-quality voice, and broadcast-quality video within a mutually optimized network.

Key: In 2000, several *Regional Bell Operating Companies* (RBOCs) reported that the volume of data traffic on their networks actually exceeded voice traffic volumes.

During the late 1990s, the seeds of change were sown that will affect the way existing and new services are delivered by content providers in the twenty-first century. The most significant changes include

- Maturity and relevance of IP routing and Ethernet LAN protocols, resulting in low-cost packet switching
- Increased demand for packet data services that now exceed demand for voice telephony services
- New, second-generation photonic components and networks, resulting in lower-cost access to bandwidth
- Development of user-friendly web browsers, resulting in new Internet services for the masses
- Worldwide deregulation of telecommunications markets, resulting in more service delivery options

Most of these changes occurred in a very short timeframe compared to the slow rollout and inertia of traditional telecom multiplexing, switching, and transmission networks. In 1993, SONET and ATM were viewed by primary telecom carriers and network integrators as the basic building blocks of all future video, voice, and data networks.

Key: By 1999, new services, service providers, and technologies, and the resurgence of old technologies changed everything. The conditions and assumptions under which SONET and ATM were expected to dominate were no longer valid.

Large enterprises (businesses) currently generate 50 percent of metro traffic. With the rapid growth of Internet-based applications and host-to-host traffic carried on private networks, many enterprises have seen their requirements for data capacity grow rapidly at a healthy 40 percent a year. These enterprises alone will require more than 20,000 OC-48 equivalents by 2005. In recent years, corporate data centers, web-hosting sites, *application service providers* (ASPs), firms specializing in storage networking and business continuity, and other network-edge players have generated explosive growth of high-bandwidth, layer 2 data traffic (Ethernet traffic). These segments alone have a 65 percent CAGR. Since this traffic requires the interconnection of backbone networks, other data centers, and access networks serving the end user, the continued growth of this segment pressures the current capacity constraints of the legacy metropolitan network (TDM/SONET-based systems).

2.2 Technology and Service Drivers: Metro Area Ethernet

Today, throughput on traditional Ethernet LANs suffers even more because users are running network-intensive software such as client-server applications, which cause hosts (workstations) to transmit more often and for longer periods of time.

2.2.1 Internet as a Driving Force

The rapid growth rate of the Internet has generated a considerable amount of publicity and is perhaps subject to just as much debate. Expectations among users, providers, and the general public are currently at an all-time high. In short, the Internet is constantly changing the face of telecommunications. Its sheer size and growing importance to business means that consistent technical standards and common practices have to be implemented on a global scale. All the functionality and capacity that will be needed for the next generation of applications must also be included in these standards. Much greater control over resources must be allowed, providers must be able to guarantee performance, and scalability must be improved.

2.2.2 Evolution of MAN Requirements

The concept of MANs is not new. They have been around since the early 1990s. In the early days, proprietary TDM rings constituted MANs and optical amplifiers accomplished the distance objectives. In the mid-1990s, ATM became the predominant technology to build MANs (ATM over SONET). The promise of ATM as the technology for converging data, voice, and video was responsible for its unanimous appeal. Most importantly, the inherent capabilities in ATM to interleave itself into the existing SONET/SDH rings made it the prime choice for transport. But a lot has happened since the mid-1990s:

- SONET/SDH infrastructures continue to be expensive and cumbersome to deploy and maintain.
- ATM has done little to enhance high-speed, packet-based connectivity.
- SONET/SDH-based bandwidth slicing is not effective to connect single users.

2.2.3 A Wish List for the MAN

Service providers deploying MANs are presented with a tremendous opportunity. LAN capacities are exploding, supported by the availability of low-cost GigE equipment. At the same time, DWDM technology has led to a huge increase in the bandwidth available in long-haul backbone networks. MAN service providers can now bridge the gap between LANs and the backbone network with new technology to support new bandwidth-intensive applications and services. Vast quantities of optical fiber are currently being deployed in metro rings. Many new products are currently being developed to harness this fiber and address the next-generation MAN market.

The following requirements for optimizing today's MAN are becoming the common list for all service providers big and small. Expanding network capacity to meet traffic demands is only one metro network market requirement. To meet future needs, metro networks must also exhibit the characteristics listed in the following sections.

2.2.3.1 Metro Core Rings The following list spells out what's required for next-generation MANs to truly be next-generation MANs:

- **Scalability** The ability to increase network capacity in a cost-effective, pay-as-you-go manner. Also, the ability to scale to bandwidths of tens of gigabits per second per link and large numbers of nodes—many more than the 16-node limit of traditional SONET or SDH systems.

- **Simplicity** Multiservice, multifunction network equipment can reduce investment and operating costs. However, some network operators are shying away from so-called *god boxes* (that is, *multiservice provisioning platforms* [MSPPs]) that portend to do everything. Delayering the network and maintaining protocol transparency are what's important in today's new MAN.

- **Support for legacy voice services** Traditional circuit-based voice is still an important revenue-generating service, although it accounts for a decreasing percentage of total bandwidth. Voice revenue, even in 2002, is still the bread-and-butter of ILEC revenue streams.

- **Powerful network management** Network management must provide extensive control and monitoring facilities yet be easy to provision and operate. The network management systems must also provide protection against misconfiguration, a cause of many network performance problems and outages. The new ideal in network management is *flow-through provisioning*, where a *graphical user interface* (GUI) allows for end-to-end automatic provisioning of network services. With a few clicks of the mouse, all involved *network elements* (NEs) are configured as required, and service is readily available.

- **Multiservice support**   Metro networks must support a wide variety of subscriber services ranging from traditional voice to broadband access (DSL and cable modems), video on demand, LAN, e-mail, and VPN. Metro networks must support many protocols including TDM, ATM, frame relay, Ethernet, and IP, and be compatible with existing SONET technology. Reducing the time to market for new services is also a key success factor.

  The network must support a range of *physical* (PHY) interfaces. It must also support service creation through extensive software-based QoS control and must support the monitoring and billing of these services through powerful *service level agreement* (SLA) monitoring tools. The most important of these services will be based on *Transmission Control Protocol / IP* (TCP/IP).

- **Robustness and reliability**   Metro networks must be able to survive multiple simultaneous faults. Protection switching, link and path restoration, and route diversity must be implemented throughout the network. Carrier-class reliability is mandatory, including 99.999 percent uptime, redundant hardware, and fiber protection and restoration capabilities.

- **QoS**   Varying levels of service must be offered for voice, ATM, frame relay, and IP-based services.

Key: Multiprotocol label switching (MPLS) may be implemented for *class of service* (CoS) offerings with IP-based applications.

- **Solid network architecture**   This architecture imitates the dual counter-rotating circuit protection capability of SONET/SDH. With IP comes layer 3 intelligence. Service providers want a robust and fast layer 3 convergence mechanism in addition to passive optical failover if a service disruption (network failure) occurs.

- **Exhaustive list of features**   In addition to the speeds and feeds that make MANs work, service providers need to pay special attention to the ability to offer tiered services. These services involve traffic classification and segregation on a per-user or per-flow basis. They also incorporate usage-based billing, policing, and authentication capabilities.

- **Greater flexibility**   The need is for interoperability with existing equipment (backward compatibility) at all locations—the CO, POP, and the customer premises. In addition, the ability to deliver bandwidth by the slice (1 Mb increments) enables service providers to offer competitive and custom services tailored to meet the needs of an increasingly demanding and savvy customer base.

- **Low cost**   The expectation of the new MAN infrastructure is a cost model that is less than SONET/SDH. This is not limited to equipment acquisition costs only. It also includes the cost involved in initial deployment and day-to-day maintenance. The cost involved in collecting data to ensure *head room* in the network (the raw bits and bytes that turn into billable services) and provisioning is where SONET/SDH has had the most negative impact on the service provider's business plan.

2.2.3.2   Metro Access Rings   The following features should be inherent in any next-generation access ring architecture:

- **Rapid and effective provisioning**   The service provider's business plan typically involves offering data-only, data/voice, or data/voice/video services. On the raw bandwidth side, the goal is to offer 1 to 10 Mbps in simple increments that are software configurable —on the fly—from a centralized management station.

- **QoS**   QoS is viewed as the only tool available to offer tiered services and rightfully so. QoS also needs to extend itself to per-user- or per-flow-based bandwidth allocation.

- **ISP flexibility**   From a MAN access perspective, it's very important to be able to connect to at least two separate ISP networks. This not only ensures an alternate path for redundancy and optimum response times for end users, but it also enables the customer to choose lower priced services for non-business-critical traffic.

- **Billing**   The most important thing of all is the ability to offer and bill for varying levels of service that include per-user billing, flat rates, always-on service, and even customized scenarios that combine flat rates and an oversubscription fee at a separate rate. No matter how well a network is structured or managed, if services cannot be billed, it's obviously all for naught.

Figure 2-4 illustrates the difference between core and access rings.

Figure 2-4
Metro (SONET) core ring and subtending access rings admit traffic from particular regional areas and route the traffic into the metro core ring. The metro core ring routes the traffic to the other region or a long-haul network. (Source: Extreme Networks "Building New Generation Metro Optical Networks")

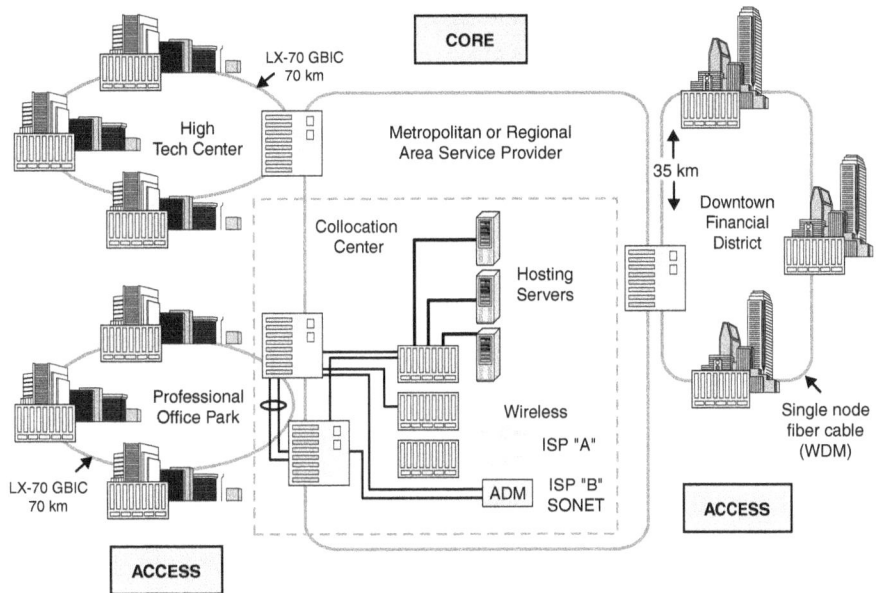

2.2.3.3 Packet Data Requirements in the New MAN The new MAN must meet the requirements of packet data services, specifically TCP/IP, in the following manner:

- MANs must be optimized for IP transport because IP will be the dominant type of traffic in next-generation MANs. Optimizing for IP transport allows flexible creation and support of new services, and minimizes equipment and operating costs by keeping the network topology flatter.

- The optimal way to transport IP traffic is over a packet-switched network with a minimum of intermediate (circuit) layers. This means lower cost, more scalability, more efficiency, and more flexibility.

- Although the MAN should be optimized for IP transport, a successful MAN solution must also support legacy transport, notably circuit voice and video (that is, ATM- and SONET-based services).

- A packet-switched network can achieve acceptable QoS guarantees only through a network operating system in which intelligent traffic planning is tightly integrated with underlying resource management mechanisms (for example, queuing engines, schedulers, and ingress traffic policing).

▪ Avoid adding any circuit-switched layers to the network. The addition of extra circuit-switched layers adds unnecessary system complexity and cost, decreases network efficiency, and makes network provisioning more complex and time consuming.

Next-generation MANs demand a solution optimized for differentiated IP-based services, most likely using *DiffServ*.

Key: DiffServ is a protocol for specifying and controlling network traffic by class so that certain types of traffic get precedence over other types of traffic. For example, voice traffic, which requires a relatively uninterrupted flow of data, might get precedence over other kinds of traffic. DiffServ is the most advanced method for managing traffic in terms of CoS. Unlike the earlier mechanisms of 802.1p tagging and *type of service* (ToS), DiffServ avoids simple priority tagging and depends on more complex policy or rule statements to determine how to forward a given network packet. A packet is given one of 64 possible forwarding behaviors known as *per-hop behaviors* (PHBs). A 6-bit field, known as the *Differentiated Services Code Point* (DSCP) in the IP header, specifies the per hop behavior for a given flow of packets. DiffServ and the CoS approach provide a way to control traffic that is more flexible and scalable than the QoS approach.

These next-generation MANs must also provide support for today's legacy voice services, which continue to provide a major source of revenue for service providers. These objectives can only be achieved through a packet-switched network that provides end-to-end service guarantees in terms of latency, jitter, bandwidth, and packet loss.

2.3   The Service Provider's Perspective

Metropolitan networks serve largely as a middleman for other networks, as illustrated in Figure 2-5.

However, this involves considerably more than providing a simple high-speed connectivity service. In fact, the value-added services and features serve to differentiate one metropolitan network provider from another. A service-rich platform should include the following features:

Metro Area Transport: 2002

Figure 2-5
Typical MAN layout

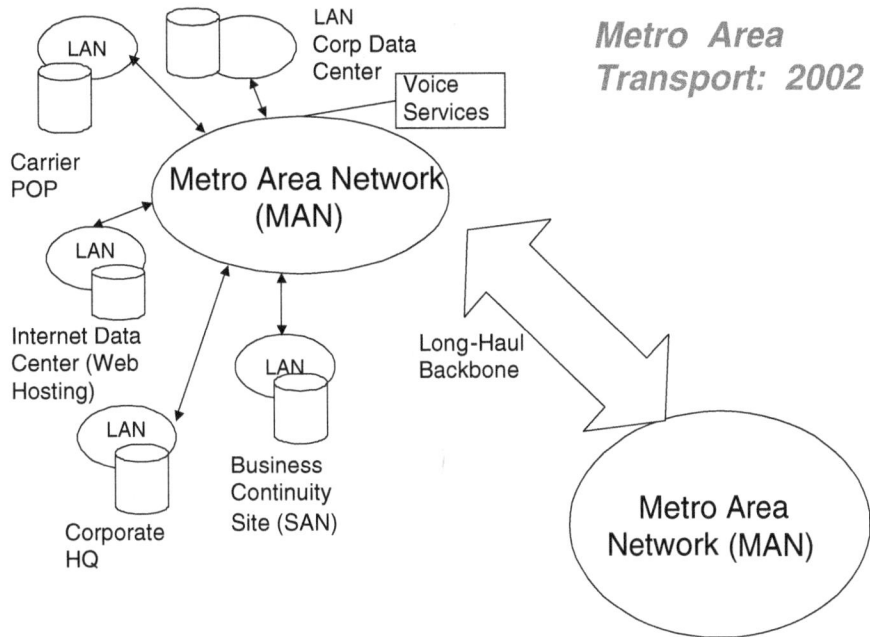

- **Leverage of the existing infrastructure** Service providers must offer advanced capabilities to support the services delivered over the metro network. For example, content and application hosting requires intelligent caching, load distribution (*server load balancing* [SLB]), and secure service partitioning without loss of performance. At the same time, access network providers require dynamic self-provisioning. Adding a new customer should not require revising the physical plant of the network. This network feature will reduce operating costs and maintain customer satisfaction. The metro infrastructure must provide security to protect customer data, reliability to route around failures, and scalability to maintain the *return on investment* (ROI) throughout hectic subscriber acquisition and growth cycles.

- **Bandwidth control and service provisioning** Service providers that make optimal use of their resources are more profitable. Bandwidth control and rapid service provisioning are among the important functions that network equipment must now perform, providing one of many opportunities for charging on the basis of various service levels. Managing data flows, muxing lower-speed data streams, and limiting network accessibility are examples of functions that manage the consumption of network bandwidth.

Key: Delivering services to the customer on short notice can be very important, especially when the service is viewed as competitive and a commodity. Today's startup Ethernet-based carriers are doing just that—enabling service changes within hours (or even minutes) via web-based portals.

The ability to set up and tear down the network links and provision optical bandwidth in bit-level increments with minimal time and effort reduces costs and retains customer loyalty. (This is in stark contrast to legacy SONET provisioning requirements.) Intelligent bandwidth provisioning and advanced traffic engineering are the catalysts for DiffServ customers. This ability also enables service providers to scale and autoprovision their optical bandwidth according to business metrics, thus creating profits by capturing revenue that would otherwise be lost.

Traffic engineering As optical networks are approaching the capability to deliver almost unlimited bandwidth to end users, service providers need to deliver DiffServ dynamically—on the fly. Various functions can be applied to the incoming traffic and traffic flows in order to improve its overall network performance. For example, service classes can be based on the identity of the customer or the type of application. Service providers can use traffic engineering to offer different service levels or ensure service quality for latency-sensitive traffic such as voice or video. Through the use of routing (such as the *Border Gateway Protocol* [BGP], *Intermediate System-Intermediate System* (ISIS), or *open shortest path first* [OSPF]), MPLS, VPNs, and DiffServ, the next generation of switch routers should be able to set priorities and offer advanced traffic engineering and hardware-based rate limiting throughout the network. These capabilities are especially critical in a world that's slowly veering away from SONET and more toward connectionless data.

Real-time accounting The transition from selling commodity optical bandwidth to selling services will require that carriers and service providers build an infrastructure that enables them to acutely manage and profit from business traffic. They must be able to monitor bandwidth usage to ensure that customers are getting what they were allocated and have paid for, and then be able to reliably account and bill for these allocations in real time. Service provider networks also need network intelligence to identify customers who consistently bump up against their limits because this presents an opportunity to sell additional services and bandwidth, and gain additional revenue.

- **Compatibility**   Older legacy networks must also be accommodated when designing and implementing new MANs to avoid redesigning local networks. As more networks migrate to IP, interoperability between IP and other telecom technologies will become a larger concern for service providers. Support should be available for any of the major infrastructure types, including ATM, *Packet over SONET* (PoS), DWDM, GigE, T1, or T3. Metro network evolution must be achieved in a way that guarantees interoperability with the vast base of already-installed LANs and WANs.

Key: Bandwidth control and accounting—not just simple bandwidth availability—lie at the heart of the new business-oriented Internet and managed network service. Without these capabilities, profitable service delivery for the metro area will be harder to attain.

2.4   Optical Ethernet in the MAN

Ethernet began as a high-speed alternative to star-wired copper in premises networks and focused on serving local applications. Over time, Ethernet has been transformed into a generic networking technology for local, campus, metropolitan, and, most recently, WANs. It has proven to be

- Scalable from 10 Mb to 10 Gb (and likely beyond)
- Flexible (multiple media, full/half duplex, and shared and switched modes)
- Easy to install
- Generally quite robust

Optical Ethernet is the use of Ethernet packets running over optical fiber within or as access to a service provider's network. The underlying connection can run at any standard Ethernet speed, such as 10/100/1,000 Mb (GigE) or even 10 Gb (as of 2002).

The research firm IDC predicts that metro Ethernet ports will increase at a CAGR of 30 percent until 2006. The factors driving this demand are similar to the factors driving other metro data services:

- Increasing corporate use of the Internet

- High-bandwidth multimedia applications

- A pressing need to connect LANs within the metro area

- A need to connect data centers, carrier hotels, and COs within metro areas

Ethernet's number-one advantage over other technologies is price. It can cost two to four times less for a customer to use Fast Ethernet or GigE in a metro area versus a comparable amount of bandwidth over private lines (such as DS-1 or DS-3).

Optical Ethernet systems are evolving beyond mere optical links that interconnect distinct LANs. They are becoming systems unto themselves, providing scale and functionality that is simply not feasible with copper-based Ethernet LANs, including those linked by routers.

2.4.1 Operational Aspects of Optical Ethernet

Key: Optical Ethernet is a fourth-generation layer 2 MAN/WAN technology. Unlike its predecessors—X.25 (first generation), frame relay (second generation), and ATM (third generation)—optical Ethernet is a connectionless packet technology. When people refer to optical Ethernet, they're usually referring to *gigabit optical Ethernet*. However, this doesn't mean that it can't also exist in 10 Mbps or 100 Mbps form.

Optical Ethernet can operate on dark fiber, optical wavelengths (lambdas), SONET, and optical rings. Along with 10/100/1,000 Mbps copper-based solutions, optical Ethernet defines a unified layer 2 campus, MAN, and WAN architecture.

Optical Ethernet can be configured on a point-to-point basis (circuit emulation or *Ethernet private line* [EPL]), a point-to-multipoint basis (emulating frame relay star networks), or on a many-to-many basis (emulating a broadcast LAN across a configured set of customer sites).

Optical Ethernet is interoperable with all legacy network protocols and architectures. That's part of its appeal. It's transparent to layer 3 transport and routing protocols, as well as to *Domain Name System* (DNS), *Dynamic Host Configuration Protocol* (DHCP), and related network tools, so it can work with legacy protocols such as the *Systems Network Architecture* (SNA) and *Internetwork Packet Exchange* (IPX). Optical Ethernet is now scalable from 10 Mb to 10 Gb speeds (via *multilink trunking* (MLT), formally known as the *Institute of Electrical and Electronics Engineers* [IEEE] 802.3ad Link Aggregation).

Optical Ethernet solutions can be privately built using dark fiber and customer-purchased *data communications equipment* (DCE), or procured as a managed service from a service provider. Many enterprises have already extended their campus networks by running Ethernet over dark fiber. Others, including Duke University and the Spring Independent School District in Texas, have deployed DWDM systems to support a combination of storage and Ethernet traffic.

2.4.2   Optical Ethernet Area Networks

Large optical Ethernet networks are changing the definition of the LAN. *Local* might even become *global*. The original limitations of an Ethernet LAN (3 km span, 1,023 nodes, and 1 optical repeater) have been obsolete for a long time. Today, the practical limits are driven by the need to terminate broadcast traffic or provide security between management domains or by today's limits on the number of *Medium Access Control* (MAC) addresses that an Ethernet switch can support. *Virtual LANs* (VLANs) have already started addressing these issues, and larger VLAN-enabled Ethernet switches in the future are at least likely to control the problems and isolate them for a high-capacity router to handle.

Today, the practical limits to the size of an optical Ethernet may not only be geographic. They involve bandwidth, node counts, and the overlying protocol in use.

Key: As VLAN and other Ethernet services become more commonplace, it's possible to envision large corporate networks simplified into a single, optically connected Ethernet LAN, with only a few large routers providing the necessary functions of security, address management, and interdomain routing.

The largest optical Ethernet networks today likely belong to carriers, interconnecting all of their POPs. Today's optical Ethernet technology could even be used to create a nationwide or global, very high-speed network that could be used to interconnect the large core routers in each POP. Such a network would have fewer router hops, faster link-failure recovery, and lower cost than today's normal network of OC-192 PoS routers. A depiction of optical Ethernet area networks is shown in Figure 2-6.

2.4.2.1 In the LAN   Today, few optical Ethernet links are implemented within a computer room or small building. However, there are exceptions for electrically noisy environments that create interference and highly secure transmissions. Even in a small building, it is easier to run a fiber-optic conduit than electrical wires because there are fewer issues with building codes.

This situation is likely to change because short-reach optics support much higher speeds than copper. It is important to note that GigE over copper is limited to a distance of about 30 m and that the next generation of Ethernet (at 10 Gb) will drop this already inadequate distance dramatically.

2.4.2.2 In the Campus Area Network (CAN)   The advent of the *Ethernet bridge*, now commonly called an *Ethernet switch*, changed the nature of Ethernet LANs. As discussed in Chapter 1, "The Fundamentals of Ethernet Technology," the purpose of an Ethernet bridge is to connect two different Ethernet LANs (the term *switched* evolved to denote interconnecting more than two LANs). Bridging occurs at the MAC layer (layer 2 in the *Open Systems Interconnect* [OSI] protocol model), and two important features are involved. First, not all traffic on either end is transported—only traffic destined for the other LAN segment is transported. Second, collisions (and collision detection signals) are not transported between LANs; each side of the bridge is basically its own LAN. Together, these features began to provide excellent enhancements to improve network performance by isolating LAN segments. They also greatly increase the maximum allowable

Figure 2-6
Optical Ethernet
Area Networks

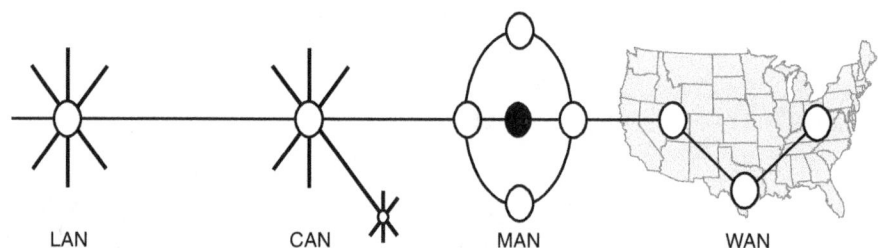

LAN   CAN   MAN   WAN

size of an Ethernet LAN. Compared to other LAN-based equipment today, bridges are not as common as they're pretty unintelligent devices.

The Ethernet bridge enabled large LANs to be deployed because a network of campus bridges could interconnect all of the building LANs together. Instead of forming a simple star network, these campus networks could be implemented as meshes, with multiple connections from LAN to LAN. This topology required the development and use of the spanning tree protocol (IEEE 802.1d), which works by disabling redundant paths (or links that have failed) and implementing a form of path protection for the LAN. As of early 2002, spanning tree protocol is still the only means to protect VLAN paths. However, the IEEE is attempting to institute enhancements to the spanning tree algorithm that will reduce network recovery time. The goal is to go from 30 to 60 seconds after a network link failure or change in link status (spanning tree parameter) to less than 10 seconds. The enhancement is called *rapid reconfiguration* or *fast spanning tree*, and it would cut down on data loss and session timeouts when large Ethernet networks recover after a topology change or a device failure.

Key: Network designers quickly realized that if both ends of an optical link terminated on a bridge port, then the traditional limits on the size of an Ethernet LAN (segment) no longer applied. The optical link could then operate in full-duplex mode, thereby doubling the effective bandwidth of the link. In addition, with only one transmitter on a LAN segment, no collisions would occur. Therefore, the need to limit the size of the span for collision detection purposes vanished.

This enabled an optical Ethernet segment to be engineered or run as far as the lasers could reach. In the early days, this still meant only a few kilometers because *light-emitting diodes* (LEDs) and *multimode fibers* (MMFs) were used, which were inherently physically limited in terms of the distances they could reach. However, this was still long enough to enable large campus LANs to be fully connected. The elimination of collisions as a flow-control mechanism required the development of a new protocol—802.3x. This full-duplex concept extended to the development of the GigE specification—802.3z.

Currently, virtually all Ethernet links greater than 600 feet (campus areas) are implemented optically. The *campus area network* (CAN) is dominated by MMF, although most CANs are really multiple LANs interconnected by repeaters or usually switches/routers that use optical links. (Some manufacturers make devices that combine the functions of a bridge and a router—these are called *brouters*.) However, this scenario is changing as the scale and functional capabilities of Ethernet switches increase. The once-dominant Ethernet 10BaseT hubs no longer offer a sufficiently lower cost to justify their use. The vast majority of LANs today are implemented with Ethernet switches, providing separate switch ports to every node (workstation) on the LAN. Traditionally, these Ethernet switches aggregate traffic into a higher-speed uplink interface (once Ethernet, then Fast Ethernet, then GigE, and now 10 GigE), which feeds a router that interconnects the LANs and provides WAN and/or Internet access.

2.4.2.3   In the MAN   Optical Ethernet in the MAN is a relatively recent development, coming of age in 1999 after the IEEE standardized GigE in 1998. Optical GigE has the capacity to provide direct, native Ethernet services as a carrier offering, with service switches that limit the actual delivered bandwidth as necessary. Multiple service providers now offer direct Ethernet services to subscribers. Only a few core routers link those subscribers to the outside world (the Internet). These providers include RBOCs and new breeds of *Competitive Local Exchange Carriers* (CLECs), which are discussed in Chapter 3, "The Metro GigE Marketplace."

Ethernet service is not the leading reason to implement an Ethernet MAN today. The desire to reduce the number of routers in the network is becoming the most compelling reason to use a layer 2 technology (Ethernet) in the metro area. The reasons are simple:

- Large-capacity routers are very expensive.
- Large-capacity routers add cost to the network.
- Every router in a circuit path adds delay to the packet transport due to the process of routing itself.

Routers usually provide an access control point to a network, maintain security using firewalls, and manage IP addresses. However, within a MAN-WAN management domain, layers of routers generate additional complexity and require several WAN administrators. The routing network is much more effective (and easier to manage) if all of the network's routers are directly interconnected, which is easily done today using optical Ethernet.

2.4.3   Advantages of Optical Ethernet

Ethernet is viewed as being media agnostic since it interfaces transparently with various transmission media including cable, copper wire, several types of optical fiber, and wireless systems. The ability to mix and match at the media level avoids significant rewiring costs that might otherwise be necessary.

Ethernet networks are becoming distance insensitive and therefore will reduce costs, simplify operations and architectures, and increase performance, all without major disruptions to existing applications.

Key: Ethernet, especially Fast Ethernet and GigE, is expected in many circles to become the leading access protocol for enterprises by 2005. By then, it's forecasted that Ethernet should account for 60 percent of total (access) bandwidth, replacing frame relay, private line, DSL, and other protocols due to the low-cost of services and familiarity in enterprise IT environments. Most of this Ethernet capacity will be supplied by the ILECs on *overbuild networks*—networks constructed and managed in parallel with legacy network infrastructures.

In summary, the benefits of optical GigE technology are as follows:

- It's scalable and delivers high performance (10 to 100 to 1,000 to 10,000 Mbps).
- It's full duplex, symmetrical, and switched.
- It has long-range distance capability—up to 150 km.
- Ethernet now has the capability to deliver service at granular rates.
- There is simplicity in network design, which reduces packet delay.
- It minimizes network infrastructure by avoiding the need for WAN protocols and related layers in the network topology, thereby lowering investment.
- Ethernet is optimized for mesh network configurations.
- In its native state, Ethernet is optimized for data-centric traffic.
- Ethernet everywhere reduces the *total cost of ownership* (TCO). People make up 35 percent of the cost of network management. The deployment of a homogenous service model such as Ethernet allows for simplicity and consistency in network management, QoS, and processes.
- The technology is stable, mature, and proven.
- Ethernet has had 26 years of evolution.
- Over 90 percent of installed enterprise networks worldwide use Ethernet (in their LANs).
- In 2000, 182,000,000 Ethernet ports shipped, which was up from 171 million in 1999.
- In 1999, 98 percent of all LAN ports shipped were Ethernet.

- Ethernet has a rapidly evolving QoS feature set.
- There is a large pool of trained network professionals.

Ethernet also has, or is developing, a wide adoption of supplementary standards:

- 802.1p for prioritization to support QoS
- 802.1q for tagging to simplify VLAN configurations
- 802.1w for rapid spanning tree, which allows for rapid failover and convergence
- 802.3ad for link aggregation, fostering scalable aggregate bandwidth

2.4.3.1 Architectural Benefits Ethernet implementations are generally done according to standards and are interoperable and interchangeable. The large amount of Ethernet suppliers has driven component prices downward and encourages continued innovation.

Another architectural advantage of Ethernet is its emerging potential to serve as a true end-to-end solution. Existing customer networks can be supported in native mode, eliminating data format conversions at the network boundaries. This removes some of the network processing that would be needed when different data link protocols are used; therefore, it reduces network complexity.

Ethernet is a scalable solution. The IEEE standards currently specify Ethernet at 10 Mb/100 Mb, 1 Gb, and now 10 Gb (June 2002). Even higher speeds are on the planning horizon and are expected to be viable in the future. Thus, network designers can start at much less than 10 GigE and build up bandwidth incrementally as capacity demand increases.

2.4.3.2 Network Management Benefits *Operations, administration, maintenance, and provisioning* (OAM&P)—the basic tasks of management systems—are also improved through the use of Ethernet across the MAN and WAN. One of the most important advantages is faster bottom-line profit through lower costs for equipment and support.

Key: Network management systems can be simplified if the same systems and technologies are used at all levels of the network.

Because new services can be remotely activated from either a provider's *network operations center* (NOC) or a customer-managed location, installation truck roll delays are eliminated. This gives customers more rapid access to the bandwidth they need.

2.4.3.3 Marketing Benefits Although it may not seem important when compared to the technical advantages, Ethernet can be easily sold to most customers. Most enterprises already have experience with Ethernet and would be willing to accept its use by service providers, especially if outsourcing is a possible option. The use of Ethernet across the MAN and WAN expands its market penetration, increasing its popularity even further. Ethernet also provides a means to avoid reengineering an existing network, which results in happier customers and network managers.

2.4.3.4 Cost of Ownership Benefits Ethernet offers measurable cost advantages because multifunction WAN routers, *data service units / channel service units* (DSUs/CSUs), *Frame Relay Access Devices* (FRADs), or ATM switches are replaced by a simple LAN switch or a router with Ethernet interfaces. *Customer premise equipment* (CPE) cost is reduced not only as a new service is installed, but also over time as existing equipment now accommodates follow-on bandwidth upgrades. Cost of ownership is further lowered as more complex, difficult-to-staff WAN administration is replaced with simple Ethernet LAN administration.

In addition, as bandwidth flexibly scales from 64 Kbps to 10 Gbps rates, customers are no longer forced to delay purchases until they can cost justify the next large step. For example, very costly migrations from DS-1 to DS-3, DS-3 to OC-3, and OC-3 to OC-12 are no longer necessary.

Key: Using optical Ethernet, customers can purchase bandwidth with the granularity they need, when they need it. This ultimately means more revenue sooner to the carrier.

2.4.4 Disadvantages of Optical Ethernet

In the interest of presenting a balanced picture, we cannot delude ourselves into thinking that optical Ethernet will be the silver bullet to solve all our network- and bandwidth-related problems in the early twenty-first century. As a matter of fact, there will be multiple metro transport technologies competing for the hearts and minds of network managers as the telecom industry continues to evolve. The following are inherent disadvantages of optical Ethernet in MANs:

- The equipment that Ethernet uses is historically designed for enterprise networks rather than public telecommunications networks. However, this is rapidly changing as Ethernet moves into carrier environments.
- Ethernet currently lacks multiservice compatibility (such as SONET, TDM, and ATM).
- It doesn't offer QoS levels required by latency-sensitive applications such as voice and video. However, its QoS toolkit is evolving rapidly to keep pace with its march into metro and WANs.
- Ethernet lacks the network management capabilities of SONET.
- It does not operate over optical ring configurations . . . yet. (See the section "Resilient Packet Ring (RPR)" in Chapter 5, "Complementary Technologies and Protocols.")
- Link and path restoration is slower than SONET protection switching. As of early 2002, Ethernet in carrier environments still makes use of the spanning tree protocol for link restoration. This equates to path switchovers of at least 1 second, while some say up to as much as 30 seconds. SONET protection switching is 50 milliseconds per standard.
- At higher port speeds, Ethernet deployment depends on the build out of metro optical fiber networks: no fiber, no GigE.
- Most Ethernet offerings on the market today are largely point-to-point solutions.
- Although GigE is well suited for moving huge amounts of data over mesh-oriented networks, it overlooks the complexities of grooming and transporting bundled voice and latency-sensitive data services. Until vendors can standardize network-wide QoS provisioning and better control jitter and delay, GigE will most likely remain in the domain of core data networks.

Not all of the changes brought on by Ethernet have everyone excited. Another drawback is that it turns routine link planning into mathematical chaos. By breaking away from the entrenched TDM/SONET hierarchy, Ethernet has no relation to the most frequently used measure of network capacity: the DS-0 equivalent. Without this standard unit of measurement, a provider that already carries a variety of traffic types cannot add Ethernet into its mix without redefining how it sizes its network.

Multiservice equipment vendors have responded to this need under the guise of trying to reclaim stranded bandwidth, not capacity planning. Nevertheless, these manufacturers have proposed using VT1.5 concatenation to bring Ethernet into the synchronous realm. Examples include running EoS or otherwise making it SONET friendly.

By providing huge access pipes, Ethernet opens up possibilities for new applications, although few of these new applications have had much success to date. Some of these potential applications hearken back to the mid-1990s video-on-demand hype; however, now the excitement is building around bandwidth on demand.

2.4.5   Key Enablers of the Development of Optical Ethernet

IDC projects that optical Ethernet will accumulate revenues totaling $741 million in 2006. Other estimates place the forecast in the low billions.

Key: When the time comes where Ethernet service is offered with RPR (by early 2004), its impact on the marketplace will be felt much more sharply.

Several factors have helped to make Ethernet a viable transport technology within metro areas. The first factor is the availability of Ethernet equipment capable of 1 Gbps speeds, namely *network interface cards* (NICs) and switch modules. Previously, Ethernet technology was limited to 100 Mbps speeds. But since 1999, GigE speeds have made Ethernet a viable alternative to private lines for linking large corporate LANs across metro areas. GigE is also used for connecting data centers, carrier hotels, and server farms within metro areas.

A second factor that has helped Ethernet to gain footing in the MAN marketplace is the build out of optical fiber networks within metropolitan areas. Although 10/100 GigE can travel over copper, optical fiber is required for 1,000 Mb (GigE) and 10 GigE speeds. Only in the last several years has a sufficient number of office buildings and other customer locations gained access to fiber facilities. However, the vast majority of buildings are still not connected.

Key: As of 2002, approximately 90 percent of Tier 1 buildings (including MTUs) are still not directly connected to fiber facilities.

Three major technology developments are key to the growth of optical Ethernet, and each can be used to link enterprises sites together or access carrier POPs, or used as a backbone technology in either private networks or service provider networks. These three technologies are 10 GigE, DWDM, and RPR:

- **10 GigE**   Ethernet's new plateau will be 10 Gbps. The IEEE 802.3ae Task Force completed the 10 GigE standard in June 2002. 10 GigE will preserve the Ethernet frame format to include minimum and maximum frame size and support for full-duplex operation. Multiple PHY optical-layer specifications are being defined for link distances appropriate for either in-building (LAN) or WAN applications, including support for 10 and 40 km spans.

  A key difference from previous versions of Ethernet is that 10 GigE will support two PHY interfaces: one optimized for the LAN (LAN PHY) and the other developed for the WAN (WAN PHY). The WAN option enables 10 GigE to be seamlessly transported across existing OC-192 SONET infrastructures. (10 GigE will be covered extensively in the section "10 GigE.")

- **DWDM**   This networking technique, which has been deployed in carrier networks (mostly long haul) for a number of years, enables multiple optical wavelengths (each with 2.5 Gbps or more capacity) to be concurrently supported on a single-fiber pair. DWDM systems can be configured in a point-to-point or ring topology. Increasingly, DWDM networks are being deployed to provide wavelength services to support applications such as storage networks for mainframe environments, and ATM and video transport.

  While to date DWDM has been delivered through optical MSPPs, DWDM interfaces are now emerging on routing switches. Enterprises will be able to implement optical architectures that support both DWDM platforms and routing switches on a single ring. In the long term, 10 GigE and end-to-end wavelength services across the WAN will make DWDM even more important.

- **RPR**   RPR is an emerging MAC (layer 2) technology that enables distributed Ethernet switching across optical rings running on fiber-optic wavelength (lambdas) or SONET. It can be implemented in routers, Ethernet switches, and optical platforms. It is being standardized by the IEEE 802.17 Working Group, and the standard is expected to be complete in 2003. However, prestandard versions of RPR are already being deployed.

  Users (*personal computers* [PCs], servers, routers, or layer 2 and 3 routing switches) can connect to RPR rings via standard Ethernet interfaces operating at 10/100/1,000 Mbps (1 Gbps). These interfaces also support IEEE 802.1q/p (see Chapter 5). Users can be locally attached if the RPR loops through the customer building or remotely attached using Ethernet over dark fiber. Figure 2-7 shows how these newer technologies can be applied. RPR and DWDM will be discussed in more detail in Chapter 5.

Figure 2-7
Metro GigE using DWDM and RPR (Source: Business Communications Review, October 2001)

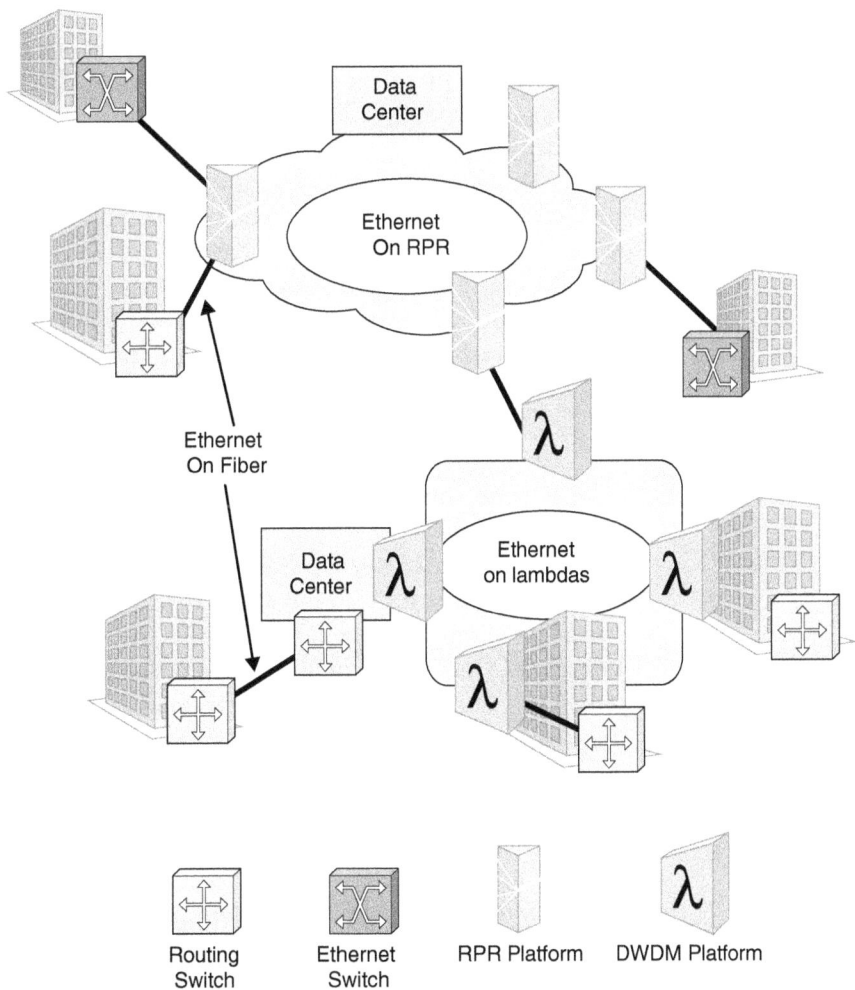

Using DWDM and RPR to Deliver Optical Ethernet

2.4.6 Carrier-Class Ethernet

With its cost leadership, ubiquity, and scale, Ethernet is the obvious choice to deliver next-generation packet services. Regardless of the underlying transport architecture (such as copper, fiber, or wireless), Ethernet as a service interface fundamentally changes the economics of data service delivery.

Consider today's scenario. Customers must deploy new, sometimes different, WAN technologies and services every time they increase bandwidth in the enterprise network. Each new technology creates added equipment cost and administrative complexity. For example, a corporate intranet (customer) expanding beyond a 2×T1 frame relay connection must invest in a fractional or full-rate DS-3 frame relay solution. The next step is OC-3 rate ATM. This same customer might also be funding hardware and service upgrades to meet Internet bandwidth growth requirements. The cost to move from 2×T1 to DS-3 is high. The cost to jump from DS-3 to OC-3 is equally huge.

Key: From the carrier's perspective, each new service install or upgrade requires a costly, often slow-to-schedule truck roll. It's not unusual for the installation to take weeks or even months. This labor-intensive process results in high operating costs, which ultimately drive service profitability down. It also introduces real revenue risk as each upgrade creates the opportunity for a competitor to claim better, faster, and cheaper services.

2.4.6.1 The Carrier Profit Opportunity   Profitability is most directly impacted by the lower operating costs that come as installation truck rolls are replaced by remote, dynamic service activation (that may include flow-through provisioning). To put these cost savings into perspective, if a carrier could eliminate even a single $650 truck roll per day across a fleet of 100 trucks, they would save over $16 million dollars over the course of one year. New-generation high-speed Ethernet switches offer a great cost reduction compared to legacy transport equipment, namely TDM gear.

2.4.6.2 Carrier Deployment Strategies and Architecture Options
The development of metro optical networks is not a simple, two-step story of rings and then mesh networks; it is a more gradual evolution. Carriers have been establishing point-to-point optical links to enterprise customers since 1999 to take advantage of immediate revenue opportunities. Star and ring topologies are facilitating a boom in network traffic and the growing demand for high-bandwidth services. However, carriers will eventually

build out fully cross-connected, optical-fiber meshes as carrier infrastructures naturally begin to emulate traditional data networks. The trend began in 2001 and will continue well into 2003 and 2004.

Point-to-Point Links The revenue potential of gigabit-speed services exceeds that of traditional leased-line services, but legacy transport equipment is not up to the challenge. Carriers' most advantageous option is adding high-speed, point-to-point optical links in parallel with their existing core legacy networks. Today, carriers can extend their metro networks when large-customer opportunities arise. One day, these *island links* can be integrated with the core when the core matures to the point when non-TDM services (GigE and DWDM) have an equal footprint in carrier infrastructures.

Stars/Rings At the onset of this phase in the development of metro optical networks, carriers are delivering many more higher-speed access services. The traffic burden on their infrastructures is mounting. Some carriers increase their transport capacity offers to other carriers and then fiber exhaustion develops in some parts of their metro networks.

As the number of customers requesting high-speed services explodes and bandwidth-on-demand services start to appear, carriers with linear-ring topologies will evolve toward full rings based on optical *add/drop multiplexer* (ADM) platforms. Carriers are beginning to see that more customers are demanding gigabit services, so it's critical for carriers to build a networking infrastructure that anticipates tomorrow's realities. When full non-SONET optical rings are eventually deployed, time to market with services will be dramatically cut and carriers will save money via reduced labor costs. Less time to deploy equals lower labor expenses.

Configurable protection capabilities will be of particular importance in the service platforms that enable this optical migration phase. A carrier will be able to determine whether an individual channel is to be fully protected on the optical layer or whether existing protection mechanisms from the SONET/SDH layer can address protection needs.

Meshes The day is coming when the logical infrastructure of optical fiber rings present in today's cities stops dictating network design. High-end users will seek flexible lambda-based services, and carriers will need to optimize the optical network for this demand. As this paradigm shift occurs, more metro networks will evolve toward mesh configurations. With the introduction of metro optical cross-connect systems, optical metro networks will be transformed into fully flexible platforms. The transformation of the traditionally voice-based network to a data-centric infrastructure will then be complete.

The most important technologies in optical-networking solutions for this phase will be cost-effective *optical-electrical-optical* (OEO) cross-connects with node-management and network-signaling software functionality similar to that of OSPF and MPLS in data networks.

2.4.7 Three Approaches to Managed Services

Three basic approaches are available for offering a managed optical Ethernet solution to enterprises: virtual private Ethernet services, services using RPR, and DWDM (wavelength services). Managed Ethernet services have the following general attributes:

- Connectivity among sites is configured on a point-to-point, point-to-multipoint, or many-to-many (multipoint-to-multipoint) basis.
- Segregation of traffic between customers and customer groups is accomplished using labeling techniques such as MPLS.
- IEEE 802.1p is used for end-to-end QoS.

2.4.7.1 Virtual Private Ethernet Services Service providers are starting to roll out Ethernet on fiber, DWDM, and RPR (prestandard versions), or a combination of the three. In this context, the demarc is essentially an Ethernet *user-network interface* (UNI). In the simplest scenario, customers directly access the POP over the appropriate Ethernet on fiber UNI (usually GigE on *single-mode fiber* [SMF]). The service provider will usually provide a demarcation device at a CP or an MTU that provides multiple 10/100/1,000 Mbps Ethernet UNIs into the network. This device is usually a high-density Ethernet switch with RJ-45 interfaces. From there, the traffic is carried over point-to-point fiber links, DWDM, or RPR optical rings. These demarcation devices enable multiple customers to share access bandwidth, especially in the MTU scenario. That said, it's apparent that a very high-capacity access pipe is in place (such as GigE and soon 10 GigE). Managed optical Ethernet services, as described here, will segregate traffic between customers and customer groups using techniques such as MPLS. IEEE 802.1p may also be used for end-to-end QoS (see Chapter 5).

2.4.7.2 Services Using RPR Today's SONET rings provide highly reliable, point-to-point DS-1, DS-3, and OC-3 connectivity for inter-router links, voice telephony trunks, legacy data connections, and ATM and video links. However, with the growth in inter-LAN traffic, SONET solutions are becoming less optimal because of SONET's foundation in TDM technology.

Key: RPR will enable service providers to evolve SONET ring networks, and RPR can be implemented by configuring OC-*n* services between user locations in a *logical* ring architecture versus a *physical* ring architecture. (See Chapter 5.)

2.4.7.3 Wavelength Services Numerous service providers have rolled out managed wavelength services for channel extension and storage applications that need to run at speeds from 45 Mbps (DS-3) to 1 Gbps (GigE). From a service provider's perspective, bit-rate and protocol-independent interfaces to DWDM systems provide faster provisioning cycles, reduced engineering times, and reduced life cycle costs.

2.4.7.4 Summary: Managed Service Approaches Most enterprises that launch an optical Ethernet implementation will begin with only a few major sites. Optical Ethernet will not become available everywhere overnight so a large number of domestic and international sites in an enterprise network will remain beyond the reach of optical Ethernet.

A key influencing factor in evaluating a migration path to optical Ethernet is the installed base of broadband capacities within the metro area(s) in question. Most sites with multiple T1s, T3s, and higher-speed interfaces into private-line and switched networks are carried over fiber. If dark fiber is available and heavily utilized, a DWDM solution may be appropriate. If a SONET ring infrastructure is in place, an RPR-based migration might be appropriate (if available). If a DWDM network has already been deployed for channel extension and storage networking, adding optical Ethernet becomes much simpler than establishing an additional network infrastructure.

Every enterprise is unique. No simple decision tree can give the right answer. Understanding the TCO of computing and networking provides important business case inputs for ultimately re-architecting information IT around optical Ethernet.

2.4.8 Enterprise Approaches to Optical Ethernet

In growing numbers, enterprises are migrating to network architectures that combine high-speed Ethernet with optical networking systems. Here are some examples:

- A major nonprofit health maintenance organization issues a *Request For Information* (RFI) for a nationwide optical Ethernet-based solution in 2001 that spans hundreds of sites.

- Almost a dozen independent school districts in Texas are rolling out optical Ethernet networks for distance learning. By centralizing routers, firewalls, and servers, they hope to free up resources for the main job: education.

- Major metropolitan school districts from Ohio to California are migrating to GigE backbones to support distance learning, video streaming, and Internet access throughout their entire districts.

Key: The justification is straightforward: simpler, faster, and more reliable networking, opportunities for rethinking server and storage distribution, and increased knowledge-worker productivity. The reason this network design shift is taking place now is due to the maturing of Ethernet transmission and switching, and the increased investment in metropolitan optical networking.

So why are early adopters migrating to optical (Gigabit) Ethernet? Here's why:

- In the networking world, Ethernet is simplicity defined. Equipment does not have to be programmed by a skilled professional in order to make Ethernet work—it's a *plug-and-play* technology.

- The capability to deploy a less expensive, high-bandwidth transport technology from LAN to MAN solves bandwidth bottlenecks easily. This is comparable to eliminating a situation where a straw is used to connect two fire hoses that are gushing bandwidth.

- Investment in installed gear is leveraged. Ethernet can take advantage of low-cost 1000BaseT NICs (such as $200 for GigE over copper).

2.4.9   Architecture Selection

Three different levels exist in any enterprise network infrastructure: access networks, metro core networks, and WANs:

- **Access networks**   Three classes of access networks are available: local access networks between a carrier POP and CP network, remote access network for off-site communications using a public network, and SANs. A local access network includes all the NEs between the provider's POP and the customer's premise network. The two major access network components are customer-owned LANs (with building, riser, and campus levels) and the MAN access link (which is usually owned by the provider).

Ethernet is the dominant technology of the enterprise LAN at both the horizontal (floor) and vertical (backbone) levels (that is, between floors and then to servers), often using a mix of the older shared media and newer switched media networks. Networks have traditionally been 10 Mb Ethernet to the desktop. However, 100 Mb and 1 Gb interfaces (NICs) are beginning to be used for high-performance users and external access.

The MAN access link is the infamous last mile between the MAN or WAN provider and the CP. The access link has traditionally been based on copper wire technologies, such as TDM, and has been relatively low in speed (especially when compared to GigE). However, aggressive fiber deployment in major metro areas and advances in last-mile access technologies have now enabled Ethernet to be used as the access link protocol.

Metro core networks   The metro core network must interface to both the local access network and the WAN access network—it's the figurative middleman in the data PSTN. Historically, metropolitan area connectivity has been provided by SONET rings, which were designed and built to carry voice traffic. However, most of the traffic growth in these networks is now due to data applications.

WANs   WANs have always been an essential ingredient in any large enterprise's network infrastructure. Various technologies have been used for low- and high-speed WAN transport including TDM (private lines), circuit switching, and packet switching (frame relay and ATM). Ethernet has only recently become a potential solution at the WAN level.

2.4.10 Summary: Optical Ethernet in the MAN

The deployment of full-duplex GigE over dark fiber using long-reach optics has eliminated the distance barriers previously associated with Ethernet technology. Use of Ethernet on an end-to-end basis is very attractive to both enterprise users and collocation providers because of its simplicity, familiarity, and relatively low cost. The recent emergence of 10 GigE creates new options for data transport over optical fiber, and will extend the value and lifespan of Ethernet technology for many years.

2.5   Gigabit Ethernet (GigE)

The 802.3z GigE standard describes multiple optical specifications for GigE. 1000BaseSX describes short wavelength (850 nm) transmission using MMF with a maximum range of 550 m on new fiber or 220 m on older fiber (with lower-quality dispersion characteristics). 1000BaseLX describes long-wavelength (1,310 nm) transmission using either MMF (with a range of 550 m) or SMF (with a range of 5,000 m). Similar to standards development for Fast Ethernet, the standards committee took advantage of existing technology and borrowed the transceivers and encoding formats of the Fibre Channel standard. The 8B/10B encoding format was used, which specifies the framing and clock-recovery mechanism. One significant change was the signaling rate, which was increased to 1.25 Gbps from Fibre Channel's 1.06 Gbps. The frame content and size are the same as previous Ethernet implementations (10BaseT and 100BaseT).

Key: Much of the early GigE standards work was geared toward specifying half-duplex operation in a collision detection environment, equivalent to the shared LANs of 10 Mb Ethernet and 100 Mb Fast Ethernet. But that work was effectively wasted because all commercial GigE implementations today are point-to-point, full-duplex links. Collision detection is *not* used in full-duplex GigE links.

Like the deployment of Fast Ethernet, the range of commercially available optical GigE interfaces exceeds the limits outlined by the IEEE 802.3z Committee. Devices that operate at 1,550 nm and ones that operate at much greater distances than 5 km are available. In fact, 150 km spans are possible without repeaters or amplifiers, depending on the equipment used. Again, like Fast Ethernet development, GigE distance capabilities have been enabled by the separation of Ethernet control logic from media control logic, which has now been formalized in a new way. It is a new standard for a *gigabit interface converter* (GBIC).

Key: A *gigabit interface converter* (GBIC) is a transceiver that converts electric currents (digital highs and lows) to optical signals and optical signals to digital electric currents. The GBIC is typically employed in fiber-optic and Ethernet systems as an interface for high-speed networking. The data transfer rate of a GBIC is 1 Gbps or more. GBIC modules enable technicians to easily configure and upgrade electro-optical communications networks. The typical GBIC transceiver is a plug-in module that is hot swappable (it can be removed and replaced without turning off the system). These devices are very economical because they eliminate the necessity of replacing entire circuit boards at the system level.

Ethernet has evolved from direct-connected routers to multiple LANs interconnected through Ethernet switches. Transmission has evolved to full duplex, with bandwidth increasing from 10 Mbps over copper to 1 Gbps, now 10 Gbps, and eventually higher over optical fiber.

2.5.1   The Rise of GigE

The 1000BaseT GigE standard was published in 1998 and is known as 802.3z. The intense growth of bandwidth demand in today's enterprise networks has ignited the drive to implement big, fat pipes on a wide scale. These new bandwidth-intensive applications have combined with the spiraling growth in enterprise network traffic to spur much higher network speeds as well as a significant increase in throughput.

In the early 1990s, before the development of switched Ethernet and Fast Ethernet (which was standardized in 1995), the industry believed that ATM (or *Fiber Distributed Data Interface* [FDDI]) would be the industry's growth path of choice. However, due to developments that began around 1994, that is not necessarily the case any more:

- The ubiquity of Ethernet has brought huge economy of scale reductions in the cost of its hardware.
- The per-port price of all Ethernet types (10 Mbps, 100 Mbps, Gigabit [1000 Mbps]) has been steadily dropping. The key: The cost factor between these different speeds is a factor of two while delivering a tenfold increase in performance.

However, even as speed increases occur in both the LAN and processor arenas, new applications have been developing even faster. They are demanding higher graphical content. New client-server applications are being implemented that require huge data transfer. Even mundane operations such as enterprise backup need huge pipes for data as memory and disk storage increase in size. All of these industry trends demand an increasing network capacity and a requirement to switch the packets at increasing speeds. GigE enables all users on a corporate network to share massive amounts of bandwidth and enables bandwidth-intensive applications such as *voice over IP* (VoIP), streaming video, off-site data management, and multimedia applications.

Key: So what is really driving the solution of today's massive bandwidth needs? GigE. With its 1,000 Mbps capacity, the 802.3z standard has now become a key element in *megabandwidth* provisioning of today's networks. Its simplicity and link to a massive embedded base of existing Ethernet technology has made it an ideal candidate for expansion into the metro area Ethernet arena.

As discussed in Chapter 1, today's Ethernet networks use a hierarchical architecture approach, with 10 and 100 Mbps switches deployed in layered arrangements, feeding into high-powered server farms and high-speed WAN routers. This topology is a natural model for the deployment of gigabit switches at the top of the network hierarchy, as GigE switches are providing a natural extension of Ethernet's functionality at the top of the enterprise network food chain that enables bandwidth-hungry applications to operate efficiently, simultaneously minimizing traffic bottlenecks.

2.5.2   User Demand and Network Strategies

Every assessment on technology starts with a statement about market demand, fueled by Internet growth, which is driving bandwidth demand. What is less clear is why enterprises are so eager to adopt new MAN technologies, even those that are untried and have no standards. Local access providers have seen this for a long time, as they saw their customers become creative when forced to make that monumental leap from a DS-1 to anything else. Such creativity included *inverse multiplexing over ATM* (IMA) and link aggregating (or trunking) multiple DS-1 lines together in order to increase bandwidth as needed.

Lower costs are being extended to the customer according to reports from carriers. Companies that had been paying between $55 and $60 per megabit are reported to be paying around $5 to $10 per megabit with GigE. Eventually, multiple price decreases look likely under a GigE price structure.

New Paradigm Research Group (NPRG) is solidly convinced that a range of delivery methods will come to coexist for a number of years into the future. The only thing known for certain is that optical fiber will not connect every customer for many years, if ever. It takes time to dig up streets or trudge through sewers. Also, the economics are just not there to connect everyone with fiber. DSL and even higher-speed algorithms are being developed to expand copper's capacity for super-fast data transport. This will fill many infrastructure gaps that exist due to geography and economics.

Key: The situation described previously is the double-edged sword of the environment of the United States. Our vast geography and capitalistic economy present us with a mixed bag. The vast geography makes it hard to fully wire the distant ends of users everywhere, but the competitive nature of our economy spurs innovation and price control by default.

There is also the matter of what applications and content will exist to justify the cost of fiber deployment with GigE when it's necessary. Many applications are available already, and new ones will be developed once an IP-based infrastructure is fully in place nationwide. The development of so-called killer applications (killer apps) will hopefully spur widespread fiber deployment. However, uncertainty surrounding MAN fiber availability could be slowing the development of these killer apps.

2.5.3   The Impact of Traffic Patterns

Due to the rapid advance of the Internet in society and the increasing development of bandwidth-hungry applications, traffic patterns have shifted dramatically. As these new applications and networks have evolved, the amount of data traffic has increased significantly, resulting in a dramatic shift in traffic patterns.

Key: The old 80–20 rule of traffic segregation has been reversed in recent years. It used to be said that network managers could expect 80 percent of enterprise traffic to remain on the LAN, whereas 20 percent of the traffic was outbound to the WAN. Today, the opposite is true: 80 percent of enterprise traffic leaves the local enterprise and 20 percent remains confined to its own LAN.

Client-server operations and the existence of VLANs have made traffic patterns much more unpredictable. The increasing use of Internet and intranet applications is putting huge demands on the available traffic routes of today's enterprise networks. What fueled the development and standardization of GigE was the comparatively high cost to upgrade and convert networks to technologies such as ATM in order to obtain the huge bandwidth boost that's been required since the late 1990s.

As processor speeds and memory sizes increase dramatically, GigE will provide the related increase in bandwidth that's required. Computers are like tripods. One leg represents *central processing unit* (CPU) power, another leg represents memory capacity, and a third leg represents network bandwidth. If any of the legs of the tripod are insufficient (too short), the fundamental computing equation becomes unbalanced. GigE restores balance to the network bandwidth part of this equation.

2.5.4   The Technical Foundation of GigE

GigE operates over the *American National Standards Institute's* (ANSI's) proven Fibre Channel PHY layer technology at 1.25 Gbps. It uses Ethernet's MAC protocols, standard *management information base* (MIB) network management format, Ethernet's frame format, and a link-level interface of 802.3. This framework means that it's compatible with all Ethernet LAN nodes already installed worldwide. Full-duplex 1000BaseLX operates at distances up to 3,000 m, whereas 1000BaseSX operates up to just 500 m. GigE also fits into the first and second layer of the OSI seven-layer model seamlessly.

Migration Options for the LAN   To take full advantage of GigE's speed in today's enterprise networks, the simplest solution is to migrate from existing 100 Mb Fast Ethernet technology. The starting point is to replace the Fast Ethernet switches that connect to the enterprise server farm with aggregating GigE switches. Another option is to install GigE blades in an existing switch that has a multigigabit backplane and then install GigE NICs in the *superservers*. These servers should have the highest possible processing speed available in order to optimize the use of GigE in the network.

Key: To emphasize the minimal impact of a migration to GigE in a LAN, note that aside from installing the new NICs and new GigE switches (or blades), nothing else in the network required an upgrade. The existing applications, desktop NICs, backbone fiber, and CAT 5 infrastructure all remain unchanged.

Another scenario where enterprise networks could be upgraded from Fast Ethernet to GigE is to replace either 10 Mb or Fast Ethernet (100 Mb) switch-to-switch backbone connections with 1,000 Mb (GigE) connections. These 1,000 Mbps connections could be supplied by either GigE buffered repeaters or 100/1,000 Mbps aggregating switches. With this huge amount of bandwidth, 1,000 Mbps (GigE) switch-to-switch links would enable the network to support a much greater number of switched Fast Ethernet connections. A third option would be to upgrade a switched Fast Ethernet backbone to a purely GigE backbone. This would enable high-performance servers to be directly connected by installing GigE NICs in the servers. With this approach, the network could support more LAN segments, more bandwidth per segment, and a greater amount of nodes per segment. In all of these scenarios, if the Fast Ethernet pipes were the only bottleneck, then substantial performance increases should be realized. Again, the beauty of this solution is that no wholesale disruption of the existing network infrastructure occurs other than the NICs and switches. Today, GigE NICs are now being installed in selective desktops that require huge bandwidth to support certain applications. Examples of these applications would be *computer-aided manufacturing* (CAM), *computer-aided engineering* (CAE), *computer-aided design* (CAD), medical imaging transfers, the transport of architectural drawings, or data mining applications (such as frequent access to a data warehouse). The secondary benefits of the GigE migration now occurring in many enterprise networks are lower training costs for network administrators, fewer required skill sets (WAN administrators), and simpler maintenance and provisioning.

Key: GigE provides a remarkable cost/performance ratio, delivering 10 times the performance for 2 to 3 times the cost.

2.5.5   Advantages and Disadvantages of GigE

On the flip side, although GigE is well suited for moving massive amounts of data over mesh-oriented networks, it overlooks the complexities of traffic grooming and transporting bundled voice and latency-sensitive data services. Until equipment vendors can standardize networkwide QoS provisioning and increase control of jitter and delay, GigE will most likely remain in the domain of core data networks.

It also eliminates the need for utilizing difficult and often inappropriate WAN technologies such as ATM or frame relay. GigE enables existing network personnel and systems to continue to grow without disruptive, radical equipment replacement or retraining of network administrators. The

Figure 2-8
North American
hardware revenue
forecast for GigE
(and 10 GigE)

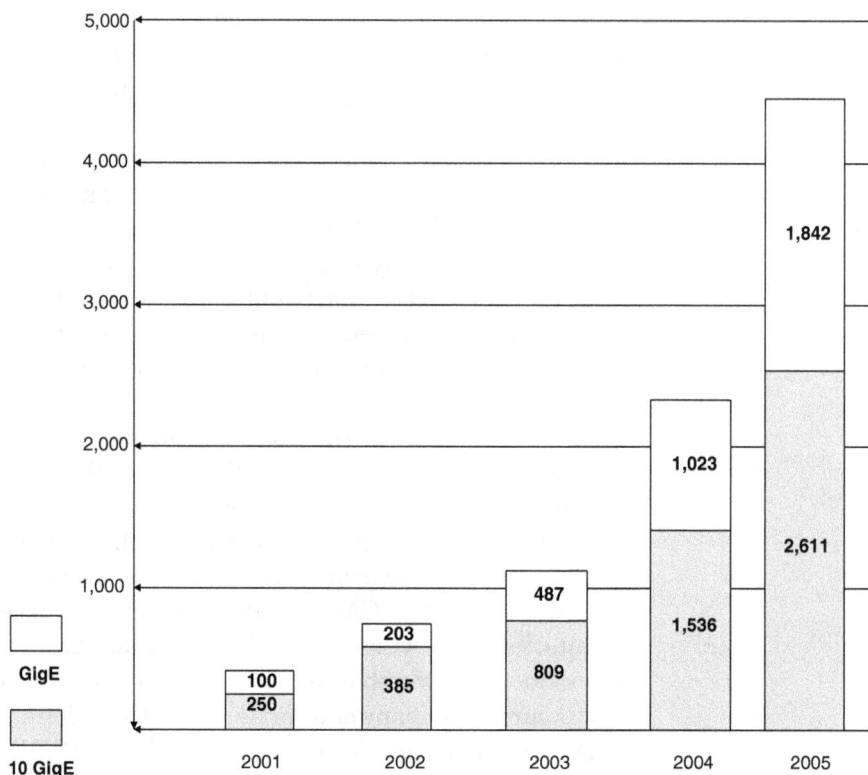

market for GigE ports sold into the carrier market is predicted to grow from
$350 million in 2001 to over $4.4 billion in 2005, as depicted in Figure 2-8.

2.6   GigE Applications

This section discusses the different ways that metro Ethernet can be used
for simple, effective applications. These applications include e-commerce
(*electronic data interchange* [EDI]), streaming video, streaming audio, connectivity to ASPs, distance learning, video conferencing, content distribution, business continuity, LAN extension (TLAN), SANs, virtual Ethernet,
and e-mail transmissions.

Key: Primary applications for optical Ethernet between 2002 and 2005
will be TLAN services, high-speed Internet access, and MAN transport.

2.6.1   Commonality in GigE Applications

The common threads among the applications for GigE and 10 GigE transport are

- **A demand for higher bandwidth**   Metro area bandwidth requirements have expanded rapidly as a result of application outsourcing, campus/building connections, and the general growth in Internet traffic. New collocation services (such as carrier hotels) require high-speed communication for data centers, application hosting, and so on. The new last-mile solutions (such as xDSL, wireless broadband, and cable modems) are also forcing metro network providers to upgrade core network capacity to keep up with their customer's use of high-bandwidth services.

- **Provisioning of feature-rich services**   Although having larger pipes is fundamental, the real key to competitive success is to create value-added services. Rapid service provisioning, priority services, and integrated management are examples of features that need to be built into an intelligent network. Service features should also be transparent and consistent across the LAN, MAN, and WAN environments.

- **Integration and convergence**   The mix and nature of traffic on the network will change as a result of convergence of voice, data, and video. Applications such as IP-based telephony and streaming video are slowly gaining acceptance, and are being combined with conventional data transfer. Converged networks generally depend heavily on the capability to manage and control the quality of the network service through the use of QoS. Convergence increases the traffic-engineering requirements for both the LAN and MAN.

- **Accessibility and connectivity**   Traffic patterns have changed from LAN centric (where the server and the clients are usually on the same LAN segment) to MAN or WAN centric. For example, the emergence of ASPs, for example, means that all their enterprise traffic must flow to an external hosting location. Connectivity no longer primarily focuses on a local workgroup; it now includes a much greater emphasis on regional networks.

- **Quality and reliability**   Network quality and reliability have become important issues for both users and service providers. Reliability can be defined for both the NEs (which can fail) and the network services (which can degrade or even become unavailable—*backhoe fade*). Most systems—both in the enterprise and at the ASP—now depend on having a network that operates within specific performance parameters on a 24×7 basis.

Key: Outages, packet losses, and degraded service become increasingly disruptive as transmission speeds increase. More data is lost, which becomes unacceptable to customers.

2.6.2   The Applications

The *Wall Street Journal* and the *New York Times* reported in front-page features on the same day in June 2001 that the fiber glut in long-haul routes stems partly from the lack of new applications that can fill long-haul pipes. Developers' reluctance to roll out new applications, in turn, comes from concern that the MAN has too little bandwidth and that not enough high-bandwidth access connections are available. In the end, the irony is that the fiber glut stems from a fiber shortage in the MAN.

The bright spot in this darkness is that the new MAN fiber is steadily being lit in conjunction with the rising popularity of GigE. This combination of metro fiber and GigE service is just a small step in dealing with the metro bottleneck, the reluctance to develop new applications, and the subsequent glut of fiber between cities.

As telecom infrastructure nationwide is built out to facilitate larger data applications, the question is just what bandwidth-hogging applications are being deployed—or not being deployed—due to bandwidth bottlenecks, whether they're real or perceived?

General Internet access, both residential and business, is surely causing a clamor for bigger bandwidth pipes. Graphic-intensive files require at least DSL-speed connections to avoid prolonged wait times. In addition, the growth in all things video—streaming video, video on demand, and teleconferencing—will probably cause a spike in capacity demand by 2004, a demand that will come from both residential and business customers. Moreover, everyone expects that e-commerce of the B2C variety will continue to drive faster connectivity. Examples include faster e-book downloading and e-catalog shopping.

2.6.2.1   Transparent LAN (TLAN) Service   TLAN service is a packet-based (Ethernet) wide area data service designed to interconnect geographically separated LANs at a data rate sufficient to carry the full bandwidth of the LAN type that's connected. Thus, a 10 Mb Ethernet TLAN service must have a 10 Mb full-duplex connection in order to carry the maximum data rate of the Ethernet in one direction. TLAN services are particularly popular since Ethernet is well suited to bridge multiple LANs within a metro on a point-to-point basis at high speeds.

Key: TLAN service enables enterprises to connect their offices within local metropolitan areas, and these sites are connected (and appear) as if they were segments on the same LAN. The goal is to remove the need for the user to upgrade to higher-end CPE via frame relay or ATM interfaces.

The customer delivers an Ethernet LAN connection to the service provider; therefore, the customer can avoid learning how to manage a frame relay or ATM network. Interconnection normally takes place at native LAN speeds: 10/100/1,000 Mbps Ethernet. Some early versions of ILEC TLAN services ran the service over their ATM backbone rather than their legacy TDM networks. Today, TLAN services are usually point-to-point, full-duplex Ethernet links.

Thousands of data center managers are now looking to outsource services that support workflow among their enterprises. TLAN service is a solution and is also known as a *virtual fiber link*. The responsibilities for implementation and maintenance of the optical networking system—and the cost and complexity inherent—reside with the carrier, not the enterprise. Enterprises spend less time and money training and deploying network support personnel and more time concentrating on their core businesses. Concerns over equipment depreciation are eliminated. Networking becomes almost invisible to the enterprise, evolving into an integrated, powerful tool that quietly enables employees to work more productively. In short, enterprises will cull greater benefits from their network infrastructures.

TLAN service is targeted to healthcare, government, education, banking and financial institutions, or any organization that must extend its corporate LAN to multiple sites in the metro. For most enterprise users, service benefits compared to private line, frame relay, or ATM are a combination of lower cost and ease of use. TLAN is without a doubt more economical than ATM or private-line solutions at DS-3 or (nxT1) rates, especially if a large number of sites are involved. Frame relay is also not generally fast enough to offer native LAN speeds. The cost savings argument is especially strong as TLAN is generally offered as a fully managed service. The end users don't need to invest in expensive CPE since the service provider owns, manages, and maintains the POP equipment, the backbone equipment, and, in some cases, even the access equipment.

Key: TLAN's superior bandwidth is finally becoming crucial. The increasing use of hosting companies and ASPs means that the LAN must be extended in metropolitan areas. The implementation of distributed *enterprise resource planning* (ERP) applications is doing to bandwidth requirements what multimedia applications were predicted to do (but still haven't really done to date).

TLAN service within MANs is seen as one of the strong early markets for optical Ethernet equipment. However, variable-rate services are likely to be more popular in the long run because few companies require high-bandwidth connections that run 24 hours a day. Figure 2-9 illustrates a sample TLAN topology in the PSTN.

2.6.2.2 High-Speed Ethernet Internet Access   After TLAN services, Internet access over Ethernet is the second largest metro Ethernet application by revenue. Ethernet-based Internet access is a shared, best-effort service supporting Internet access that scales from 64 Kbps to 1 Gbps with fixed-rate and burstable-rate options. Service providers can generate increased revenue with significantly reduced cost and service complexity. This service offers retail Internet access to businesses, schools, and government agencies over metro Ethernet networks. Much of the Ethernet-based IP access business is displacing T1 and DSL Internet access.

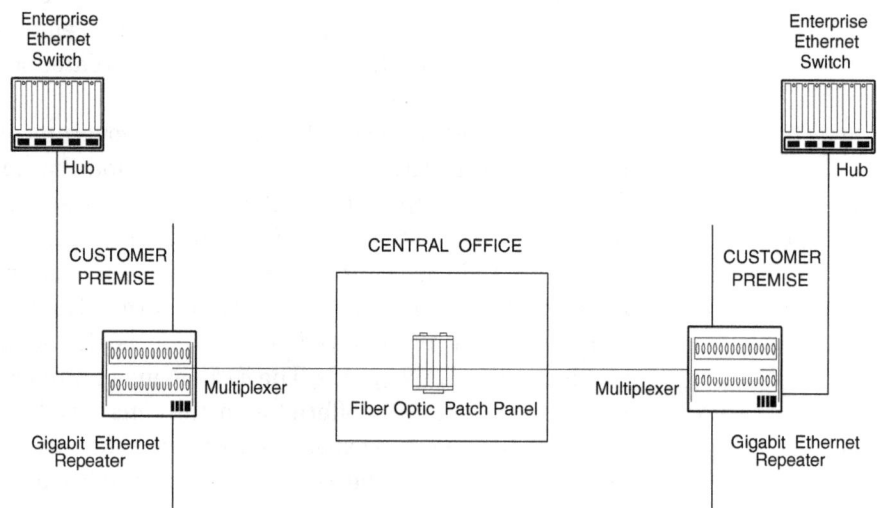

Figure 2-9
TLAN topology

2.6.2.3 Inter-POP Connectivity Another large market segment for optical Ethernet will be within service provider networks, where it can be used to create the carriers' local backbone, interconnecting cable modem networks or DSL networks. This means interconnecting data centers, collocation facilities, multiservice operators, and carrier POPs. This effectively makes this application a form of TLAN service.

The Yankee Group predicts a significant rise in this type of service as GigE services continue to catch on due to reduced cost, simpler management, centralized network resources, and the possibly greater security of key databases and storage servers. Similarly, secure VLANs (sort of an inter-LAN VPN) may also catch on in the metro, provided the security is truly up to par and the provisioning meets the customers' flexibility requirements. Vendors such as Lantern Communications say they can support thousands of VLANs on a single ring, which simplifies the carrier's bandwidth allocation and prioritization duties.

GigE actually extends the LAN into the MAN and WAN. Therefore, any LAN service should be able to be duplicated across the WAN as well. This enables applications such as hosting, disaster recovery, storage backup, and streaming media to exist over a WAN. GigE's flexibility is also a godsend to the struggling ASP market, which partially blames its customer churn on the need for a more flexible high-speed network—a network that responds quickly and cost effectively to changing customer usage patterns. Today's dedicated lines cannot match GigE's capability to provision rapidly and in granular increments (1 Mb at a time).

2.6.2.4 Storage Area Networks (SANs) For many businesses and ISPs today, telecommunications and availability of data are strategic business factors that are as important to a company as labor and financial strength. Telecom infrastructures and company data run mission-critical applications such as billing and accounting on mainframe computers or powerful client-server installations in data centers. These mission-critical applications simply cannot afford extended periods of nonavailability (outages). However, these applications require real-time transfers of increasing data volumes that, in some cases, already exceed several terabytes.

Key: Continuous backups or reliable decentralized data processing and storage have become essential components of any major business infrastructure. This need has become even more important since the attacks on the United States in September 2001.

A SAN is a high-speed special-purpose network (or subnetwork) that interconnects different kinds of data storage devices with data servers on behalf of a large network of users. A SAN is usually part of the overall network of computing resources for an enterprise. It is usually clustered in close proximity to other computing resources such as IBM S/390 mainframe computers. (In a computer system, a cluster is a group of servers and other resources that act like a single system and enable high availability and in some cases, load balancing and parallel processing.) SANs support disk mirroring, backup and restore, archival and retrieval of archived data, data migration from one storage device to another, and the sharing of data among different servers in a network. SANs have generally consisted of a Fibre Channel network linked to servers and to a range of storage devices such as disk arrays or tape libraries.

Today most SANs also extend to remote locations for backup and archival storage, using WAN carrier technologies such as ATM or SONET, and now GigE. More recent SAN designs use GigE instead of Fibre Channel for all the same reasons that GigE is catching on in other areas of the enterprise and public networks: simplicity, seamless extension of LANs, less network equipment to buy and manage, and a reduced number of protocols and technologies to deploy and manage.

The popularity of SANs has risen so much in the early twenty-first century that several companies' sole business is to outsource the management and operation of SANs for other companies. These SAN providers will lease or build their own data centers and work with their customers to provide connectivity to the data centers using leased (and sometimes owned) facilities.

Centrally managed storage schemes have become more economical, and demand for broadband connectivity has risen accordingly. This is especially true at the MAN level, where the majority of SAN connectivity occurs.

The relentless advance of e-commerce and other data applications is driving a need for data warehousing, causing an increase in demand for storage. Much of the demand for storage will be distributed storage; therefore, it will require high-speed connectivity. This high-speed requirement stems from simple physics—more territory equals more transport distance. More transport distance equals more potential latency. Higher-speed compensates for potentially higher latencies.

2.6.2.5 Ethernet Private Line (EPL) An EPL is a point-to-point, dedicated bandwidth service where bandwidth can be software provisioned in granular increments that scale from 1 Mbps to 1 Gbps. In leveraging the low cost and scale of Ethernet, this solution creates a compelling lower TCO migration strategy for (nxT1), DS-3, and higher-rate private-line services. EPL will replace standard TDM-based dedicated or leased lines with—you guessed it—Ethernet. The trick is to ultimately combine Ethernet's quick and granular bandwidth provisioning with SONET's reliability and security (not exactly Ethernet's strong points). The result should be much lower-cost, fat-pipe bandwidth on demand.

Even the incumbent telcos are launching EPL services. SBC's GigaMAN product is one such offering. EPL is very attractive to incumbent carriers if it can provide the same level of security as traditional private lines. It's an improvement over traditional T1 services because it can be bandwidth adjusted very easily through software and offers a very low cost of ownership for the customer. Such a service could disrupt incumbent-dedicated line prices. The Yankee Group predicts EPL pricing in Tier 1 metros may have to drop by 40 to 50 percent to stay competitive. For EPL to take off, it must offer SLAs across the WAN, not just in the metro. For instance, best-effort Internet over the WAN provides no QoS today once traffic passes beyond the metro. For the time being, that means using SONET may be best in some cases because it's ubiquitous. If EPL evolves to where it has a global (or near global) reach, customers may come running.

2.6.2.6 Ethernet Virtual Private Lines This service (or application) is a flexible, fully SLA-guaranteed, point-to-point or multipoint service that is designed to enable many users to share a single path for greater bandwidth efficiency and lower cost of ownership. Applications might include broadly deployed TLAN services and more cost-effective frame relay access.

Key: The services listed previously can extend beyond the metro area to the wide area by leveraging the SONET/SDH infrastructure to achieve global scalability, end-to-end manageability, and resilience. This will become even more possible with the approval of 10 GigE in 2002. Carriers will also gain the benefits of low cost of ownership and faster time to market by integrating these optical Ethernet services into their existing (legacy) global infrastructure.

2.6.2.7 Ethernet Virtual Private Networks (EVPNs) A slightly sexier transmission service on the horizon is *Ethernet virtual private network* (EVPN). Karen Barton, Vice President of Marketing at Appian, defines EVPN as "a multipoint-to-multipoint service that can be either fully secure [a premium service] or shared [lower cost] for enterprise customers, industry consortiums, collaborative communities, and other closed-user groups who have three or more metro area sites requiring high-speed connectivity."

This service would differ from TLAN in that a big pipe can be securely shared across the appropriate user nodes. Each user node can use Ethernet's flexible bandwidth provisioning for a distinct SLA-level CoS or on-demand access to that same pool of secure bandwidth for truly tactical service. There's one catch though—each node must support Ethernet. Still, that shouldn't be a showstopper for most industry segments.

2.6.2.8   Virtual Private Ethernets   An important selling point for Ethernet is the security it offers through the use of VLANs. VLANs associate individual Ethernet switch ports with a specific LAN and split broadcast domains into smaller, easier-to-manage sub-LANs. But what service providers really like is that VLANs can run over trunks between switches by associating tags (unique labels) with ports. When extended to a MAN, users can be on the same VLAN as colleagues across a campus or city, and the carrier can create a SLA that guarantees that no hacker or unauthorized party will join the group, because the carrier manages the switch port(s) that supports the VLAN. Equipment vendor marketing departments have turned this once ordinary technology into an exciting new concept called *virtual private Ethernets*.

2.6.2.9   Virtual Ethernet Rings   Conceptually, a virtual Ethernet ring extends the advantages of a VLAN across a wide area SONET infrastructure to include broadcast, discovery, and multipoint capabilities. Based on the capability to statistically share a SONET path across multiple users, the virtual Ethernet ring can be offered at any explicit bandwidth rate, scaling from 1 Mbps to 1 Gbps. A customer could also be given a premium service that is built as a physically secure, dedicated time slot through the network, much like a standard private line or VPN tunnel. This topology offers

- The ability to securely segregate one customer's traffic from another customer's traffic
- The ability to deliver explicit bandwidth guarantees
- WAN-extended RPR services (RPR standards are expected in 2003)
- Standard SONET infrastructure and benefits between locations
- Centralized, software-based service provisioning

A virtual Ethernet ring is provisioned as a series of physical SONET paths that interconnect network sites in a logical ring configuration. A provisioning tool that could be integrated into a carrier's existing operating system would greatly simplify and automate this process. Once the path is established, all carriers have to do is define each member of the virtual Ethernet ring via port and service identification.

Each customer's corporate site enters the ring through a standard Ethernet interface. This makes it easy to remotely modify or change the amount of bandwidth available. With a virtual Ethernet ring, customers are given their own private Ethernet broadcast domain that makes full use of Ethernet's multipoint capabilities. They are no longer required to buy an expensive mesh of private-line circuits to link sites. Instead, they buy a ring that is actually constructed as a series of shorter SONET paths that enable full, reliable connectivity across the WAN. This offering is essentially point-to-point EoS.

2.6.2.10 Telemedicine Teleradiology is another specialized application that relies on fiber-type speeds.

Key: Teleradiology is a heavy user of bandwidth because by law x-rays cannot be compressed. The typical medial image file such as an x-ray, CAT scan, or *magnetic resonance imaging* (MRI) is 50MB in size. This automatically means more bandwidth is required to transport medical imaging files.

Although print and broadcast graphics producers can compress their files, their image files regularly exceed hundreds of megabytes if not a gigabyte, so compression can only do so much to reduce the need for bandwidth. What adds to the problem is the need for simultaneous channels that can carry video communications between medical or graphics professionals when they need to discuss their transmitted material.

All such applications will require a considerable amount of bandwidth at the metro level, especially over the last feet. Any point of congestion across the network will hamper efforts to roll out such services, suppressing revenues all around, along with workplace productivity. Even the perception of too little bandwidth will affect an application developer's decision to create and produce new telecom and IT applications. It's likely it already has.

2.6.3 Summary: GigE Applications

Figure 2-10 sums up the applications and market segments that are the key drivers of today's GigE applications. Broadband networks—GigE—are at the center of today's applications.

Figure 2-10
Application and market drivers for GigE in the metro network

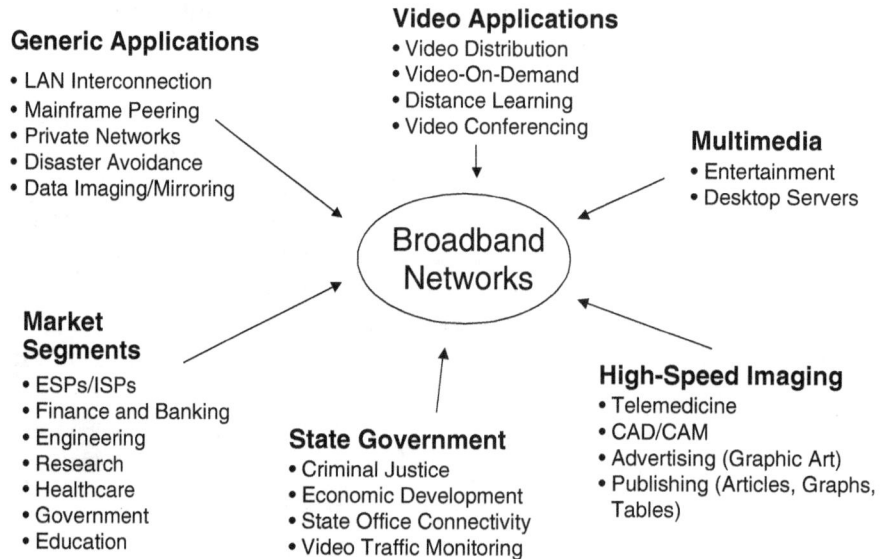

Generic Applications

• LAN Interconnection
• Mainframe Peering
• Private Networks
• Disaster Avoidance
• Data Imaging/Mirroring

Video Applications
• Video Distribution
• Video-On-Demand
• Distance Learning
• Video Conferencing

Multimedia
• Entertainment
• Desktop Servers

Broadband Networks

Market Segments

• ESPs/ISPs
• Finance and Banking
• Engineering
• Research
• Healthcare
• Government
• Education

State Government
• Criminal Justice
• Economic Development
• State Office Connectivity
• Video Traffic Monitoring

High-Speed Imaging
• Telemedicine
• CAD/CAM
• Advertising (Graphic Art)
• Publishing (Articles, Graphs, Tables)

2.7   10 GigE

Although GigE was just introduced in 1998, some customers are already clamoring for faster technologies. The average IT shop probably isn't feeling the squeeze on its 1 Gbps backbone yet, but vendors are still preparing for the next level of speed and performance. 10 GigE is well beyond the concept stage, and although there might not be a need for 10 billion bits per second of capacity today, it would behoove enterprises to build their network infrastructures to support this technology in the near future.

Many IT shops only recently migrated to GigE technology in their network backbones, and the idea of yet another technology migration may seem daunting. Although some experts predict that network spending will drop sharply well into 2003, the evolution of technology and applications will still drive the need for more bandwidth, especially in the network core.

The traffic load on a network is directly proportional to the amount and type of systems attached. As enterprises continue to deploy GigE links toward the network's edge (the users), the demand for more bandwidth at the core will inevitably increase.

2.7.1   Why 10 GigE?

Why all the fuss about 10 GigE? Because GigE has done so well. Because Ethernet, at all speeds, has a proven track record. As the dominant LAN protocol, Ethernet is well understood with very few integration problems and no special CPE requirements. Ethernet prices per port are declining 30 percent annually, making it an incredibly cost-effective technology. 10 GigE can be used with or without SONET or ATM, dropping infrastructure prices even further, which minimizes network complexity. Ethernet technology has not only survived, but it has thrived because of the ease with which new technologies have been able to gain more bandwidth from the same infrastructure. Ethernet has done it again, multiplying its capacity tenfold.

One way to gauge just how excited we should get over 10 GigE is to look at the projected size of the market. At the higher end, Lantern Communications claims that the metro market for 10 GigE will hit $2 billion by 2004. According to the Gartner Group, 10 GigE service should be available in 100 cities by the end of 2003, as opposed to the roughly 30 or so cities where it was available in late 2001. 10 GigE revenue is expected to grow from around $200 million in 2001 to $3.5 billion in 2004, per Gartner DataQuest. See Figure 2-11 for a graphic that reflects this growth trend.

Pricing will be key to market acceptance, and adoption depends heavily on whether customers can get fiber to their buildings, as 10 GigE will only operate over glass.

Key: Currently, only about 10 percent of buildings in the United States have direct access to optical fiber facilities. It can be very expensive to construct fiber laterals to buildings, and this issue might raise prices in the long run. The amount of money MTU customers are willing to spend on Internet access is also a key issue.

So what makes it a worthwhile technology?

- It will scale enterprise and service provider LAN backbones. In other words, it provides the means to easily migrate to the next highest speed in the Ethernet technology hierarchy (10, 100, 1,000, and 10,000 Mbps [10 GigE]).

- Similar to GigE, it enables you to extend Ethernet to both the MAN and WAN, providing transparent data connectivity between the LAN, MAN, and WAN.

Figure 2-11
10 GigE revenue
projections—2001 to
2004

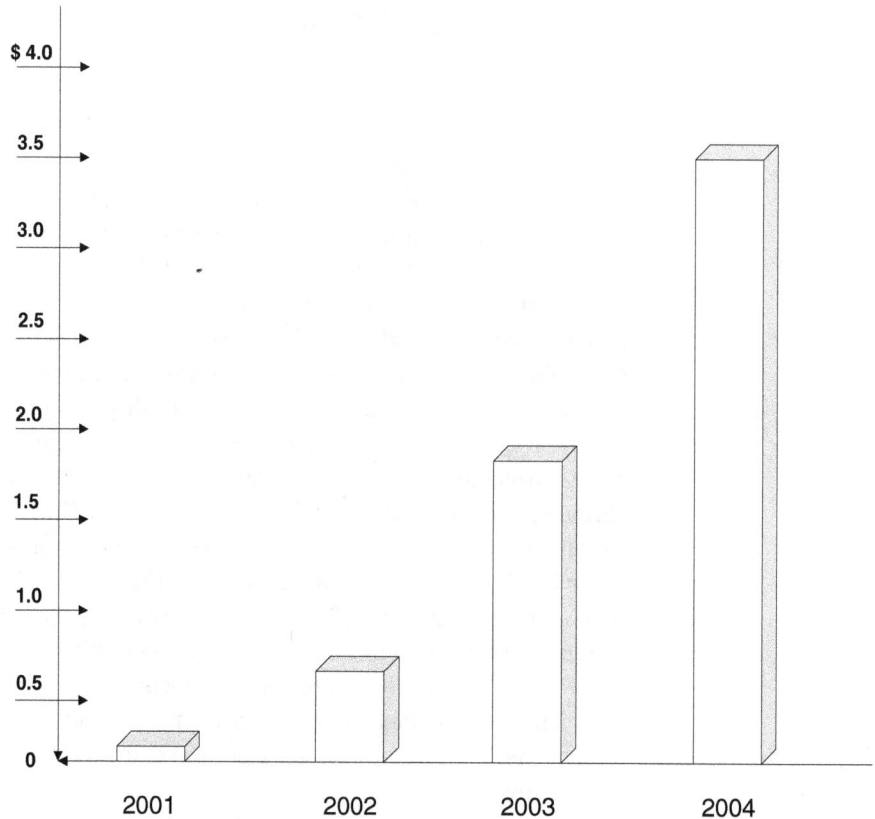

- 10 GigE makes use of the Ethernet installed base—a base of more than 300 million Ethernet ports worldwide.

- With 10 GigE, all traffic types can be supported—data, voice, and VoIP.

- It is faster, simpler, and more cost effective than other alternatives. It's the best in TCO when compared to other protocols and technologies.

Another major driver for 10 GigE in the enterprise is the proliferation and widespread adoption of GigE over copper in the wiring closet. Wiring desktops for GigE may seem like overkill, but soon desktop systems will have integrated 10/100/1000BaseT (LAN on motherboard) NICs. Apple Computer's G4 tower computers now come with integrated 1000BaseT interfaces as standard. Prestandard products were available in late 2001 from startups Atrica and Lantern Communications, more established players like Extreme Networks, Riverstone, and Foundry, and the perennial powerhouse Cisco Systems. Although the average desktop user may not need 10,000 Mbps (10 GigE) all day, he or she will benefit significantly from decreased wait times on large file transfers when they are required. This type of use will create very bursty traffic patterns on wiring-closet uplinks, ultimately increasing backbone network load. In this context, 10 GigE implementation is a clear solution for the enterprise backbone.

Key: It's not unrealistic to believe that GigE will be a common technology at many desktops by 2005.

It's still unclear how the market for 10 GigE equipment will develop. Most attention has focused on the fact that this will be the first version of Ethernet aimed at both service providers (via the WAN PHY portion of the standard) and enterprises (via the LAN PHY part of the standard). However, although 10 GigE will likely be a simple, logical next step for enterprise networks, the service provider market looks a little more complicated. Nevertheless, 10 GigE will extend the success of its remarkable predecessors, but at faster speeds. The answer is not black and white, and a lot of uncertainty still remains about specifics in terms of SONET and voice support, deployment methods, and QoS.

2.7.2   Enterprise Market Drivers

Almost everyone agrees that the earliest adopters of 10 GigE will be very large enterprises with an immediate need for higher-capacity aggregation of GigE links. GigE aggregation will be a key driver of early 10 GigE deployments. Aggregation could occur in three places in large enterprise networks:

- In building basements, aggregating Gigabit links that come through the risers from wiring closets
- In *storage area networks* (SANs)
- At server farms

At these aggregation points, 10 GigE should be a clear winner. Today, the only option high-bandwidth users have is to trunk multiple 1 GigE ports together between two switches using *multi-link trunking* (MLT) to create a backbone of 2 Gbps or higher. This technique is useful to attain higher capacity and as a means to provide failover for network redundancy.

However, MLT has downsides. For one thing, it uses up valuable Ethernet switch ports. The second and bigger problem is that it may not always deliver the capacity users expect. According to sources at Spirient, some tests have shown that because of the way the control plane and data plane interact in trunking scenarios, on some Ethernet switches, there is hardly any additional bandwidth gleaned when using MLT.

2.7.3 Market Forecasts

One way to gauge just how excited we should get over 10 GigE is to look at the projected size of the market. Estimates vary pretty widely. At the higher end, Lantern Communications claims that the metro market for 10 GigE will hit $2 billion by 2004. According to the Gartner Group, 10 GigE service should be available in 100 cities by 2003, as opposed to the roughly 30 or so cities where it was available in late 2001. However, this doesn't mean that service is available to a particular building unless fiber extends directly to the CP. This explains lower market penetration estimates. Communications Industry Researchers' assessment of 10 GigE only estimates an $800 million market in the United States by 2005. Regulated telcos (the RBOCs) probably won't enter the market until 2003. The early adopters will likely be major metro MTUs. The lack of availability of voice services related to 10 GigE may eventually inhibit market growth as well.

2.7.4 10 GigE Standards Development

The following characteristics were the criteria used by the IEEE 802.3ae Committee when standardizing 10 GigE:

- **Technical feasibility**   The 10 GigE standard had to be technically feasible using existing proven technology. Reliability should be at least equal to earlier standards.

- **Standards conformance**   Ethernet networks that operate at different speeds need to be compatible (10 GigE has to be Ethernet in more than name only) in order for scalability to be preserved across a network. For example, 10 GigE needed to be compatible with GigE, GigE needed to be compatible with Fast Ethernet, and Fast Ethernet needed to be compatible with 10BaseT. Most of the benefits of end-to-end Ethernet would be invalid if gateways between segments were necessary.

■ **Broad market potential**   10 GigE has to provide distinct benefits over competing solutions (such as SONET OC-192c and ATM). Since Ethernet access links at 100 Mb and 1 Gb are now being used, core networks must be able to aggregate multiple Ethernet access links (Fast Ethernet and GigE). The number of vendors that are committed to producing standards-based 10 GigE products reflects a definite market appeal.

■ **Economic feasibility**   The costs associated with producing new 10 GigE products, implementing the solutions, and maintaining the implemented 10 GigE networks should be reasonable when compared to other versions of Ethernet and other competing technologies. Costs should be reasonable for the performance expected.

Key: In order for 10 GigE to look like traditional Ethernet to its users, the MAC user interface must be the same.

2.7.4.1   The 10 GigE Standard   The 802.3ae Standards Committee was formed in January 2000 at the recommendation of an IEEE study group to develop a standard for 10 GigE. The standard was approved in June 2002. Unlike Ethernet standards of the past, the 10 GigE standard will no longer support shared media access. At that speed, it's impractical anyway. It is highly unlikely that any significant work will be done on copper standards for 10 GigE in the near term. However, it should be noted that in January 2002, Ahmet Tuncay, CEO of a startup called SolarFlare, claimed that the firm was working on a standard that enables 10 GigE to run over copper.

2.7.4.2   10 Gigabit Ethernet Alliance (10 GEA)   Supplementing the work of the 802.3ae Standards Committee were the efforts of the *10 Gigabit Ethernet Alliance* (10 GEA), an industry consortium of about 100 equipment vendors working to promote the acceptance and success of 10 GigE. The group is not the same as the IEEE 802.3ae Standards Committee, which worked on the actual standards for 10 GigE. The equipment manufacturers didn't passively wait on a standard because several vendors wanted to announce 10 GigE products during 2001, in advance of the standards ratification expected in 2002. Of course, each vendor hopes that their products will reflect (one of) the standards, and they promise that they will conform once those standard interfaces are defined. Table 2-2 shows key aspects of 10 GigE links as defined by the 802.3ae Task Force.

Table 2-2

Key aspects of 10 GigE links as defined by the 802.3ae Task Force

10 GigE Links
850 nm serial   • 65 m on existing MMF   • Up to 300 m on new MMF
1,310 nm *wideband wave division multiplexing* (WWDM)/serial   • Up to 300 m on MMF   • Up to 10 km on SMF
1,550 nm serial   • Up to 40 km on SMF

2.7.4.3 Distance Pairings   The 802.3ae Task Force has defined several distance goals and has issued preliminary suggestions for minimum distances over various types of fiber. The task force identified five key distance pairings: two pairings for MMF and three pairings for SMF. Figure 2-12 also depicts 10 GigE's relationship to the OSI reference model.

For the LAN, the task force has suggested that the 850 nm serial *physical media dependent* (PMD) interface should support a working distance of 65 m on 500 MHz 50/125 micron MMF. This type of fiber is not prevalent in today's installations. However, because of its low cost and ease of manufacturing, it may become the interface of choice for interserver connectivity at 10 Gbps. The 1,310 nm WWDM interface should support at least 300 m on 62.5/125 micron MMF.

Key: 62.5/125 micron MMF is the most common type of MMF installed in the enterprise. It is also known as FDDI-grade fiber and is typical of fiber installed in the late 1980s and early 1990s. Depending on the given fiber installations and desired end-to-end distances, enterprises may be able to use some of their existing fiber infrastructure when migrating to 10 GigE.

For longer-haul CAN, MAN, and WAN applications, the 10 GigE Task Force recommends a 1,310 nm serial or WWDM interface that supports distances of at least 10 km (~6 miles) over SMF or a 1,550 nm serial interface that supports at least 40 km (~24 miles) over SMF.

Figure 2-12
10 GigE and the
OSI model

| MAC Layer | MAC (Media Access Control) Full Duplex Only | | | |
	Optional XGMII (10 Gigabit Media Independent Interface)			
Physical Layer	1550 nm PHY	1310 nm WDDM PHY	650 nm PHY	WAN PHY Encoding 1550 nm PHY 1310 nm PHY

2.7.4.4 Media-Independent Interfaces Because of all these PHY interfaces, the task force has also introduced the possibility of media-independent interfaces, similar to the *attachment unit interface* (AUI) of 10BaseT Ethernet. Although several standards for a universal interface have been proposed because of the sensitive electrical and timing issues involved with 10 GigE, we most likely will never see a truly vendor-independent interface. Instead, some vendors may have a 10 Gb media-independent interface proprietary to their own product suite. Table 2-3 lists fiber types and the distances each type supports for those interfaces that are actually media dependent.

Prestandard products started appearing on the market from nearly all existing providers of high-speed layer 2 and 3 switching products during late 2001. At that time, 10 GigE was being deployed mostly in data centers. It was expected that as the standard neared completion, prices would reflect the typical incremental costs of previous versions of Ethernet (that is, 10BaseT and 100BaseT). In December 2001, multimode 10 GigE ports cost approximately $30,000 to $60,000 and single-mode pricing was in the $15,000 to $20,000 range.

Key: Prices for 10 GigE equipment will drop by at least 50 percent during the first year of availability (2002 to 2003) and be followed by the historically proven Ethernet price decreases of 30 percent year over year.

The 10 GigE standard is significantly more complex than the previous Ethernet standards. When 100BaseTX technology was introduced in 1993, vendors were able to leverage the FDDI physical interface (copper) to facilitate development of the standard. When GigE was introduced in 1998, vendors leveraged the optical and electronic interfaces developed for Fibre Channel. However, 10 GigE does not have a comparable technology to leverage. To develop a 10 Gb PHY layer, vendors had to develop a layer capable of transporting 10 billion bits of information per second!

Table 2-3

Fiber distance specifications

Optic Type	Fiber Supported	Fiber Diameter (Microns)	Maximum Distance Allowed
850 nm serial	Multimode	50.0	65 m
1,310 nm (WWDM)	Multimode	62.5	300 m
1,310 nm (WWDM)	Single mode	9.0	10 km (~6 miles)
1,310 nm serial	Single mode	9.0	10 km (~6 miles)
1,550 nm serial	Single mode	9.0	40 km (~24 miles)

The IEEE 802.3ae 10 GigE Task Force outlined several key objectives that should be of interest to IT managers. The most important is that 10 GigE will remain an Ethernet technology. In other words, it will preserve the 802.3 Ethernet frame format, minimum and maximum frame size, and MAC client service interface. Also, like the GigE standard, the 10 GigE standard will be a full-duplex-only standard. Like the GigE 802.3z standard, *Carrier Sense Multiple Access with Collision Detection* (CSMA/CD) will not be included in the 10 GigE standard. The GigE standard included support for a special half-duplex repeater-based implementation called a *buffered distributor*. However, this technology never made it into widespread deployment. As a result, 10 GigE will not include a half-duplex option. This decision not only solves the problem of network diameter (switched networks have no inherent distance limits other than cabling limitations), but it will also significantly simplify the standard. Half duplex was once implemented for cost reasons since it was cheaper than full duplex. However, given that most of the cost of gigabit and 10 Gb interface is in the optics, half-duplex connections no longer make sense. They aren't and won't be available.

2.7.4.5 Physical Media Dependent (PMD) Interfaces PMD specifications were the most difficult and contentious part of developing the new standard. In an effort to accommodate the various parties' interests regarding cost, distance limitations, and physical plant, the committee decided from the start to include four distinct PMDs in the standard. Table 2-4 describes these four PMD standards.

Out of necessity, the committee chose a set of PMDs to handle the various link lengths and trade-offs of cost versus serving as large an applications base as possible. However, virtually all of the PMD devices are new. The PMD problems caused the committee to miss its March 2002 deadline.

Table 2-4

10 GigE PMDs

PMD (Optical Transceiver)	Type of Fiber Supported	Target Distance in Meters
850 nm serial	Multimode	65
1,310 nm WWDM	Multimode Single mode	300 10,000
1,310 nm serial	Single mode	10,000
1,550 nm serial	Single mode	40,000

Four options were studied for 10 GigE, which were all distinguished by distance. Vendors brought prestandard versions to market with the goal of tweaking those versions to be compatible with the 10 GigE standard after it was ratified. Different applications for 10 GigE technology will require different versions of the standard—different PMDs. The first, least expensive solution is the short haul, which is less than 65 m, using 850 nm serial optics. The second option is 1,310 nm *wide wave division multiplexing* (WWDM) , which has three distances: 100 m (MMF), 300 m (MMF), and 10 km (SMF). A third option is 1,310 nm serial, which provides maxiumum distance of 10 km (~6 miles) on SMF only. Last, there is 1,550 nm, which provides distances up to 40 km.

2.7.5   Two Flavors of 10 GigE

The 10 GigE standard will define two PHY layers: the LAN PHY and WAN PHY. 10 GigE is designed to support not only enterprise backbone networks (LAN PHY), but also wide area and long-haul networking applications (WAN PHY). These options will allow for two types of Ethernet-based carrier services. The first will be MAN services running at native Ethernet speeds, probably constructed with a mesh of high-speed layer 3 switches. The second offering will be WAN based, with rate adaptation to OC-192 speeds. By using SONET framing and existing DWDM capabilities, the distance limitations of standard Ethernet links will be eliminated.

Key: The LAN PHY represents the typical LAN interfaces used today, operating at a 10 Gbps data rate. The WAN PHY is designed to operate at a data rate compatible with the payload rate of OC-192c and SDH VC-4-64c

(9.58 Gbps). The 10 GigE WAN PHY will operate at 9.952 Gbps to enable Ethernet to map directly to the WAN SONET and DWDM infrastructures. However, unlike SONET networks, 10 Gb EoS will remain asynchronous. This will enable 10 Gb to run over SONET networks, but the switches and routers will not need the complex and expensive stratum clocks that native SONET interfaces require.

2.7.6   Media Models for 10 GigE

The following sections detail how 10 GigE will operate over different networks and transmission systems.

2.7.6.1 LAN/MAN Media   A number of alternative media types are being considered for LAN/MAN environments. Since 10 GigE is not expected to connect directly to user and systems (at least not yet), the standards are initially being restricted to optical fiber. Options being considered for the optical layer include MMF and SMF, using both serial and parallel links.

2.7.6.2 WAN Media   The concept of Ethernet-based communications over long distances would not be feasible using the original CSMA/CD protocol (that is, shared media with contention). The restriction to full-duplex operation over optical fiber links enables 10 GigE to operate over long link spans, using repeaters and other transport layers such as DWDM or SONET. *This approach was taken with GigE as well.*

The 10 GigE WAN PHY layer is also compatible with SONET. 10 GigE will be SONET friendly even though it is not fully compliant with all of the SONET standards.

Key: Some SONET features in the 10 GigE standard still need to be implemented, such as the OC-192 link speed, the use of SONET framing, and some overhead processing. However, the most costly aspects of SONET will be avoided, such as TDM support, performance requirements, and management requirements.

Fiber is the physical media of choice for 10 GigE not only because of increased distance compared to copper cable, but also because of its scalability through the addition of WDM.

2.7.7   GigE and 10 GigE

Table 2-5 provides a brief comparison between the functional and standards-based differences between the 1 and 10 Gb versions of Ethernet. Several important restrictions apply to 10 GigE. The traditional CSMA/CD protocol will not be used at all, even in LAN PHY operations, and copper wiring will not be an option, at least not for end station links.

Figure 2-13 illustrates in detail all the interface and distance relationships with GigE and 10 GigE, and 1000Base-N specifications.

2.7.8   Getting Ready for 10 GigE

Whether a business or carrier is interested in 10 GigE now or planning for the future, several concerns must be addressed. If they are planning to retrofit a building or a campus, the challenge lies in adapting the infrastructure to support the new data rates.

FDDI-grade cabling will support a maximum distance of just 300 m. For many campus environments, this means that MMF will no longer be suitable for backbone networking when using 10 GigE.

Single-mode LAN installations will be available for ~6 to 24 mile configurations, with the potential for 48 miles or more as laser technology continues to improve. If the campus does not have sufficient SMF to replace

Table 2-5

*GigE attributes
versus 10 GigE*

Attribute	GigE	10 GigE
Physical media	SMF and MMF, WWDM (802.3z), and twisted-pair copper (802.3ab)	SMF and MMF
Distance	Up to 5 km (~3.1 miles)	Up to 40 km (~24.8 miles)
MAC	Full and half duplex	Full duplex only
PMD	850 nm, 1,300 nm, and 1,550 nm WWDM	850 nm, 1,310 nm, and 1,550 nm
Accommodates OC-192 TDM and WDM	No	Yes
Packet size	Variable 64–1,514 bytes	64–1,518 bytes

Figure 2-13
802.3ae/z/ab
distance chart

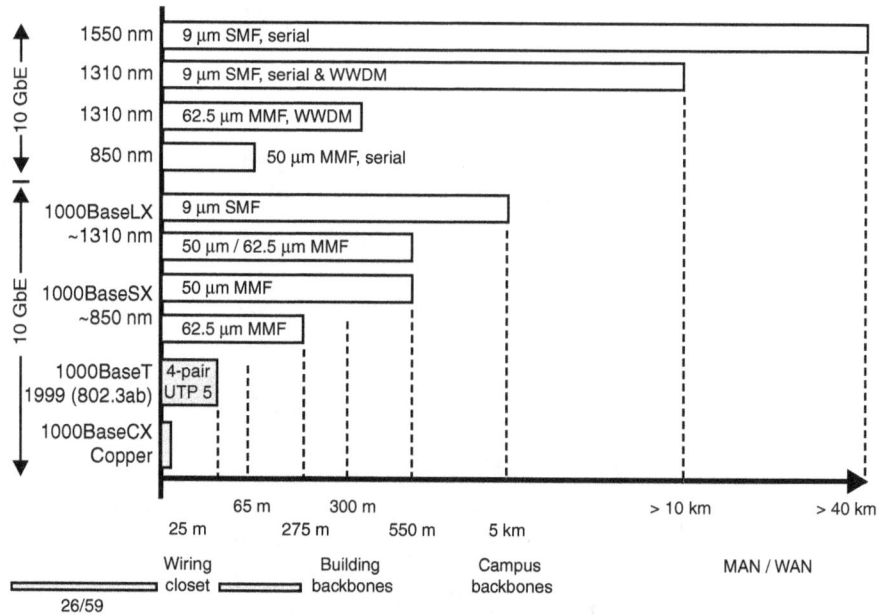

10 GbE ← 10 GbE →	1550 nm	9 μm SMF, serial
	1310 nm	9 μm SMF, serial & WWDM
	1310 nm	62.5 μm MMF, WWDM
	850 nm	50 μm MMF, serial
10 GbE	1000BaseLX ~1310 nm	9 μm SMF
		50 μm / 62.5 μm MMF
	1000BaseSX ~850 nm	50 μm MMF
		62.5 μm MMF
	1000BaseT 1999 (802.3ab)	4-pair UTP 5
	1000BaseCX Copper	

25 m 65 m 275 m 300 m 550 m 5 km > 10 km > 40 km

Wiring closet   Building backbones   Campus backbones   MAN / WAN

26/59

existing multimode runs, network managers should upgrade now if planning to run GigE campus-wide. When renovating campuses or when buildings are under construction, it's necessary to provision a sufficient amount of SMF to support the new requirements of 10 GigE. SMF was once an expensive option compared to multimode installations, but network managers should consider single mode not only for interbuilding links, but for vertical risers as well to plan for the future. Futureproofing buildings is the critical objective when designing and implementing structured cabling systems.

Key: 10 GigE is expected to cost three to five times more than today's GigE (1000BaseX) interfaces when it's available. Early modules from Extreme Networks and Foundry Networks cost $40,000 or more, but a $10,000 to $15,000 initial price range is more likely now that the standard is ratified.

The good news is that many existing vendor architectures, including Foundry's BigIron, Extreme's Black Diamond, and Cisco's Catalyst 6500, will support 10 GigE modules in their current form. Although these architectures were only marginally designed for 10 GigE, they can still use the technology to significantly improve data rates compared to 1 GigE, thereby protecting existing investments. Cisco Systems and other vendors already offer 24-port 10/100/1000BaseT gigabit-over-copper modules for around $375 per port. In 1995, 10/100 Fast Ethernet interfaces cost about the same, so it's easy to see how Ethernet scales and simultaneously becomes cheaper over time.

2.7.9 10 GigE Applications

The reemergence of MANs is being stimulated by the growth in e-commerce and the outsourcing of key enterprise services such as extranets, back-office automation, telecommuting, content/web hosting, and application services. In the early days of the Internet (1994 to 1996), a simple, low-bandwidth link offering a best-effort CoS was readily available and easily used, but these low speeds are no longer sufficient.

Advanced applications include video streaming, voice telephony, and online music distribution. These applications depend on continuous access to a broadband network infrastructure to operate efficiently—meaning not having to wait forever for processing to complete. These applications can no longer be considered bleeding edge. Today's business-critical applications depend on the network being available at all times, especially with transaction-based systems such as online purchasing, application hosting, and soft product distribution. Applications that reside on a service provider's host or that are collocated with a network backbone provider also depend on high-quality intelligent networks. Examples of these new classes of application are listed in Table 2-6.

Dark-fiber metro applications—that is, running 10 GigE over leased dark fiber between campus buildings in metropolitan areas—will likely be the first applications to use 10 GigE. This will enable high-speed connectivity across metro regions for a fraction of the cost of leased lines.

The protocols and techniques for bandwidth sharing over parallel links exist, work well, and are used in thousands of sites. It is a simple step to run parallel optical Ethernet trunks, each on a separate wavelength, all multiplexed over a single-fiber pair using DWDM technology. This way, a point-to-point Ethernet link could have scores of 10 Gb channels, with an aggregate Ethernet bandwidth of perhaps 400 Gb. For example, by using recently announced DWDM capacity of 160 wavelengths, 1,600 GigE links could be implemented. So in a sense, Terabit Ethernet is already available. Of course, this kind of network would require very high-capacity Ethernet switches at the ends of that fiber.

Table 2-6

Emerging 10 GigE
applications

Application Area	General Requirements and Description	Application Name
High-speed Internet access	The aggregate Internet bandwidth requirements increase as the number of Internet users grows. Also, the amount of time each user spends on the Internet is also increasing due to new applications that are being deployed. Newer applications typically also require more bandwidth and, in many cases, higher QoS.	10 GigE LAN PHY Internet access
Corporate LAN interconnection	LANs must be interconnected for distributed communications, remote services, and home office access. Since Ethernet dominates the LAN environment, seamless connectivity among geographically distributed Ethernet segments is highly desirable.	TLAN service
Back-end server connections	Servers that were once distributed to the department level have now been consolidated into more central server farms that must support high transaction rates. Server networks need high bandwidth to minimize congestion and delay.	10 GigE LAN PHY
Inter- and intra-POP connections	The service provider's POP interconnects the local access and core metro layers and may also include access to the server farms. The POP itself could include multiple networked switches to enhance scalability and reliability, and minimize congestion.	10 GigE LAN PHY
Real-time streaming	Real-time data transfer places demands on the network that can often be offset by using higher capacities. Internet radio, VoIP, and video on demand are real-time communications applications. For example, video data streams are by nature high-bandwidth and jitter sensitive. This type of data transport (either broadcast or on demand) is only practical when a broadband infrastructure is available. Metro networks that provide the bridge between high-speed WANs and LANs must not become a performance bottleneck. An example of this application is the video streaming that's available on www.cnn.com.	10 GigE WAN PHY

Table 2-6 cont.

Emerging 10 GigE
applications

Application Area	General Requirements and Description	Application Name
Telecommuting	A home office network should be capable of operating at the same speeds as the office LAN and provide a transparent connection between the two LANs. Provision of metro or regional Ethernet connectivity supports an office LAN that extends to the home with virtually no degradation in service.	TLAN via remote access
High-speed data transport	Various applications are emerging that would only be acceptable when large amounts of data can be transferred while a person waits (in other words, almost in real time). Software downloading (such as www.real.com for Real Player) and music distribution are just two examples of these new applications.	10 GigE LAN or WAN PHY (depends on end-to-end distance)

Source: "Building 10 Gigabit/DWDM Metro Area Networks," techguide.com

2.7.9.1 10 GigE in the LAN   The first applications for 10 GigE in the LAN are mundane; they interconnect high-speed equipment in the POP or collocation center. The Yankee Group expects the first metro application to be the aggregation and back haul of 1 GigE traffic. The catalyst in both cases is to slash costs for interconnecting 1 GigE interfaces.

Key: Although the metro-core-oriented Ethernet players will be challenged to provide the quality and reliability associated with SONET, the limiting factor on the penetration of Ethernet as an access technology is the availability of last-mile fiber. Although 100 and 1,000 Mbps Ethernet (GigE) can run over copper, there are distance limitations of 328 feet ($\sim$ 100 m). (1,000 Mb ethernet over copper is very distance limited—that is, it's used in *campus area network* [CAN] applications only.)

2.7.9.2 10 GigE in the MAN   10 GigE will change the way enterprises use their networks. It will also change the face of MANs and WANs. With LAN PHYs and SONET-compatible WAN PHYs, Ethernet will take on a whole new role in the MAN. The applications for 10 GigE range from the obvious expansion of LAN backbones to MAN and WAN services. In the

LAN, 10 GigE will be used primarily as a switch-to-switch link in large backbone networks, running through vertical risers in collapsed backbone LANs. It will also play a smaller role as an aggregator between workgroup switches where GigE is justified at the desktop. 10 GigE will not be found in end systems/servers until *input/output* (I/O) technology improves to the point of a higher-capacity architecture such as InfiniBand.

InfiniBand is an architecture and specification for data flow between processors and I/O devices that promises greater bandwidth and almost unlimited expandability in tomorrow's computer systems. Between 2002 and 2004, InfiniBand is expected to gradually replace the existing *Peripheral Component Interconnect* (PCI) shared-bus approach used in most of today's PCs and servers. Offering capacity of up to 2.5 Gb and support for up to 64,000 addressable devices, the architecture also promises increased reliability, better sharing of data between clustered processors, and built-in security. InfiniBand is the result of the merging of two competing designs: Future I/O, developed by Compaq, IBM, and Hewlett-Packard; and *Next-Generation I/O*, developed by Intel, Microsoft, and Sun Microsystems. For a short time before the group came up with a new name, InfiniBand was called *System I/O*. Like the migration to GigE, enterprises running very large campus backbones must ensure that their switch platforms have the necessary backplane and buffering capacity to adequately support 10 GigE links.

Other LAN-oriented applications for 10 GigE will be for *serverless* buildings, also known as "serverless backup," as well as business continuity. Serverless buildings are those locations where data backup doesn't occur on a server. With server-free backup, data flows into a SAN directly from a (server) disk drive (PC) to a tape device, with no data moving through the server whatsoever. The enterprise servers only need to host the backup application. Server-free backup removes a large hurdle in the drive to optimize enterprise network performance. It enables companies to fully leverage their investment in SAN technology. Serverless backup is depicted in Figure 2-14.

Key: The importance of developments like InfiniBand in terms of their impact on 10 GigE is that as devices at the edge of today's networks (such as PCs and servers) migrate to newer, faster technologies, the rest of the network must keep pace in order to avoid developing network bottlenecks.

10 GigE won't replace SONET by 2008, but the decline of traditional SONET will begin by that time. Some estimates place pure Ethernet services at 30 percent or more of the metro market infrastructure by 2005.

Figure 2-14
Serverless backup

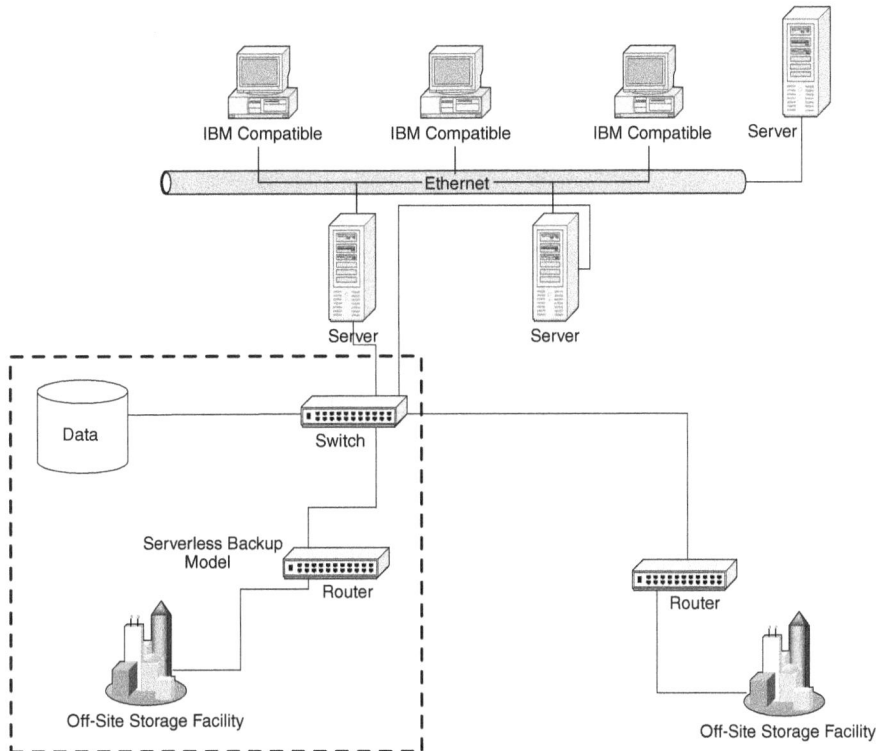

IBM Compatible   IBM Compatible   IBM Compatible   Server

Ethernet

Server   Server

Data

Switch

Serverless Backup
Model

Router

Off-Site Storage Facility

Router

Off-Site Storage Facility

2.7.9.2.1 GigE in the MAN over DWDM The need to expand metropolitan network capacity to accommodate high-speed WANs is well recognized, but finding innovative, cost-effective solutions has not been easy. Many metropolitan network service providers have been deploying SONET and WDM across existing fiber despite the difficulties in supporting existing customer networks, particularly Ethernet. A new solution is emerging— 10 GigE over DWDM—which offers the advantages of speed, flexibility, scalability, and technical simplicity. Extensions to existing Ethernet standards have proven to be both feasible to develop and practical to implement. The 10 GigE standard is now complete and will be used for open system deployments.

Many metro network providers now need to determine how they can incorporate 10 GigE into their network infrastructure. As the amount of 100 Mb Ethernet links at the edge of their networks increases, so will the need for 10 GigE to aggregate 1 Gb links to data centers and in the core of the network.

2.7.9.2.2 *The Advantages of Running 10 GigE over DWDM* The advantages of a combined 10 GigE/DWDM solution can be examined from three distinctly different perspectives:

- The advantages of each technology by itself
- The benefits of having Ethernet as a service
- The operational and managerial advantages of end-to-end consistency

Each is important to consider when evaluating solutions to offer to customers.

2.7.9.3 **10 GigE in the WAN** Ethernet transport has not yet taken off in the long-haul network, but this is expected to change as 10 GigE interfaces become available in 2002. Some of these interfaces are expected to operate at SONET OC-192 speeds and at the distances required for long-haul networks. The distance limitations are not a serious concern because most long-haul networks use DWDM to combine multiple circuits over a single fiber in a daisy-chain fashion. Each circuit is on its own wavelength, and these DWDM systems provide the long-haul capability themselves due to their inherent optical amplification feature.

> *Key:* By using SONET framing for WAN PHY connections and existing DWDM capabilities, the distance limitations of standard Ethernet links are eliminated.

The main drivers of wide area Ethernet deployment are speed, low cost, and simplicity. For instance, consider a nationwide 10 Gb IP/SONET-style ring implemented as OC-192 PoS links between a dozen cities. Each city would need a large router with two relatively expensive PoS ports, and the average packet would traverse half a dozen routers as it crossed the network. In the event of a link failure, the routers would spend a significant span of time converging on a new set of routing tables to bypass the failure.

Now consider the same network with each city containing an Ethernet switch with two 10 GigE switch ports and a 1 Gb port connected to the local router. Total costs here would be much lower because 10 GigE switch ports should be much less expensive than the equivalent router ports and 1 Gb router ports are already relatively inexpensive. A 10 GigE WAN port will likely range from $10,000 to $50,000, whereas PoS interfaces are around $295,000. Each router sees a direct connection to every other router, which simplifies table lookups. In the event of a link failure, the Ethernet switches reroute the traffic more quickly at layer 2—the routers do not need to be involved. Speed. Low cost. Simplicity. These 10 GigE architecture options will dramatically shift the price-to-performance curve of WAN services.

Key: New Ethernet-based services are already offering an increase in price to performance by 2.5 to 25 times.

WAN services using 10 GigE will offer significant savings over traditional solutions. By directly coupling new Ethernet capabilities with a more robust optical core, service providers will be able to eliminate SONET aggregation and transport layers from the network topology.

Now that 10 GigE has been standardized (in June 2002), the IEEE has recognized Ethernet's potential to be used as both a MAN and WAN technology. With Ethernet's historical limitations stripped away through the adoption of a full-duplex, fully switched architecture, it can now provide access services into the WAN.

The WAN PHY standard will be a SONET OC-192-compatible stream, which is clocked at 9.92 Gbps. The disadvantage is that it is not exactly 10 times 100 Mb Ethernet. There may also be cost disadvantages—SONET is not the least expensive transport method. However, the advantage is that it would be guaranteed to interoperate with all OC-192 SONET devices, including the networks of all of the major telephone companies, and all of the OC-192-compatible DWDM systems. The reality is that 9.92 Gbps is close enough to 10,000—no real applications are likely to notice the difference anyway.

The *WAN Interface Sublayer* (WIS) inserts 10 GigE into a SONET STS-192c frame by adjusting the interpacket gap to accommodate the slightly lower data rate of OC-192c payloads (9.92 Gbps). Table 2-7 shows a timeline of the development of Ethernet *speeds and feeds*, along with a data transfer perspective.

Although the 10 GigE WAN standard will include SONET framing and a subset of the SONET overhead, it will not meet the same jitter requirements.

Table 2-7

History of Ethernet speeds and feeds

Year	Technology	Speed	Data Transfer per Hour
1972	Ethernet	10 Mbps	3.6 Gb
1993	Fast Ethernet	100 Mbps	36 Gb
1998	GigE	1,000 Mbps	360 Gb
2002	10 GigE	10,000 Mbps	3.6 Tb

Jitter is the deviation in or displacement of some aspect of the pulses in a high-frequency digital signal. As the name suggests, jitter can be thought of as shaky pulses. The deviation can be in terms of amplitude, phase timing, or the width of the signal pulse. Among the causes of jitter are *electromagnetic interference* (EMI) and crosstalk with other signals. Jitter can cause a display monitor to flicker, affect the ability of the processor in a PC to perform as intended, introduce clicks or other undesired effects in audio signals, and cause loss of transmitted data between network devices. The amount of allowable jitter depends to a major degree on the application in question.

10 GigE switches will not require costly stratum clocks for timing and synchronization. That alone will drive down costs to deploy 10 GigE, but it also raises the question of whether or not 10 GigE can ultimately deliver voice and video services (without a stratum clock for timing). Operating in the metro domain, latency will not be an issue since properly engineered rings are less than 40 km (24.86 miles) in length.

2.7.9.4   The Magnitude of a Killer App   No one's sure exactly which end-user apps will push 10 GigE to mass acceptance, but there is no shortage of network uses, especially in the metro—for example, collocation and storage network connectivity. The 10 GigE standard will support Fibre Channel as well as inter-POP connectivity, especially for back-haul purposes and aggregating 10/100 Mbps Ethernet streams and router-to-router traffic. Service providers can also hand off traffic to upstream providers more efficiently and cheaply by using a single 10 GigE pipe. However, major impact to business models depends on prices dropping significantly, at least in the metro. 10 GigE's first impact in carrier businesses should be in wholesale services.

Network applications will impact end-user behavior down the road. According to Bob Walters, Executive Director of Optical Networking at SBC, "10 GigE has such enormous bandwidth that service providers and end users could use it for literally any application. Some examples are SAN connectivity, data-center disaster-recovery and mirroring systems, multimedia transport, CAD and medical file transport, and data-center access."

2.7.10   10 GigE Pricing

Key: The 802.3ae Committee's goal for 10 GigE is to deliver 10 times the speed of GigE at 3 to 5 times the cost.

Despite improving prices at the 1000BaseT (copper) level, the average selling price for GigE remains in the neighborhood of $1,000 per port or higher, according to Extreme's Duncan Potter, who estimated that the average 10 GigE LAN port would cost between $8,000 and $12,000 by early 2003.

Bruce Tolley of Cisco concedes that PMDs are driving up the cost of the early implementations, but he's convinced that as purchasing volume increases, component prices will fall. In the long run, having a customer base that includes both enterprises and service providers will help generate volume quicker. "That's why there's a lot of interest in the 1,310 nm serial PMD that works on up to 10 km of SMF, claimed Tolley. That transceiver can work in an enterprise environment for uplinks and over a dark-fiber metro network. It can also be used as an interface into DWDM equipment. So the 1,310 nm serial PMD has the ability to grow a large market quickly because it addresses multiple application areas—not just the enterprise, but also the service provider space."

Once the pricing lowers sufficiently from a projected $50,000 per port (when the first switching blades ship), 10 GigE will prove itself to be cost effective in metro networks and data centers.

Key: It's estimated that Ethernet in the wide area may be one-fifth the cost of SONET and one-tenth the cost of ATM. Ethernet is also rightfully viewed as being media agnostic, since it interfaces transparently with various transmission media including copper cable, wireless systems, and several types of fiber.

2.7.11 Deployment Plans

Given their traffic volumes, carriers' requirements for high-capacity equipment are even more pressing than enterprises'. Ethernet promises improvements in provisioning and service delivery costs, along with higher throughput.

Where will 10 GigE demand occur? The service provider market is going to develop faster than the enterprise market because service providers recognize that speed is ultimately a competitive distinction. A traditional Fortune 500 company has to ask if 10 GigE will give them an advantage. It could simplify some aspects of network deployment, but in terms of normal day-to-day operations, 10 GigE is not going to be something they'll jump on immediately.

Despite this optimism, there may be equally good reasons why 10 GigE will move into the carrier networks slowly. For one thing, the incumbent carriers have always moved slowly when making their purchase decisions, even before the capital expenditure crunch of 2001 and 2002. However, there's also the matter of culture and embedded base.

Key: Enterprises are fully comfortable with Ethernet and their networks are loaded with the technology. By contrast, incumbent carriers are still getting used to the idea of Ethernet in their service networks.

2.7.12   Is It Really This Simple?

Realistic benefits from the deployment of 10 GigE should be realized by 2004 if the economy doesn't have another downturn. So what are the problem spots to watch out for when choosing a 10 GigE service? Selling 10 GigE to upper management is one issue. Like GigE, 10 GigE will also require the following:

- Standardization of reaches (short haul versus long haul)
- DWDM interoperability
- QoS that includes bandwidth management for voice and private-line applications
- SONET-type fault tolerance and restoration capabilities
- Protocol and traffic agnosticism—the ability to carry any type of traffic without requiring format conversions

Although 10 GigE slashes costs by not having to deploy expensive SONET techniques for synchronization and timing, on the flip side, it raises real-time QoS issues. Those issues remain a major concern in some carriers' minds. "QoS will be tough to administer with 10 GigE unless somehow it can be deployed over IPv6, which is still a long way from being ready to deploy on a widespread basis," says SBC's Bob Walters. "If 10 GigE is deployed over DWDM, some type of protection may be possible using 'digital wrapper' technology, but that is also bleeding edge and will need to be tested thoroughly prior to being made available to customers."

However, the vendor community has an answer to the QoS and protection dilemma: RPR technology, a layer 2 protocol designed to bring SONET-type capabilities to the data world. Led by Nortel, Cisco, Luminous

Networks, Dynarc, and Lantern Communications, RPR is a work in progress, and standards are due in 2003. RPR builds on the concept of packet rings, derived from Cisco's proprietary *Dynamic Packet Transport* (DPT) protocol.

Aside from the lack of robust standards and compliant chipsets for vendor gear, the two biggest impediments to a rapid rollout of 10 GigE are ultimately

- The lack of compelling service applications to drive ROI
- The cost of deploying fiber to unconnected buildings

Research with retail-focused *Ethernet service providers* (ESPs) indicates that metro fiber deals with wholesalers such as MFN, Qwest, Level 3, and Williams have not provided access to enough on-net buildings. (Author's note: As of June 2002, Williams has declared Chapter 11 bankruptcy; and Qwest and Level 3 are in dire financial straits due to the telecom sector economic funk of 2001 and 2002.) This is because these fiber providers target incumbent providers' COs and hosting facilities instead, per another Yankee Group report. The result is that the ELECs must build costly lateral connections off of the metro core fiber rings to connect to target customer buildings, at an average cost of $75 to $100 per foot, depending on geography. Whether or not the ESPs will consider this depends on their business model and the potential for *return on investment* (ROI).

In spite of all the caveats, 10 GigE is still a good bet for the MAN and will eventually be one for the WAN. Given the RPR initiative, ELEC's assumed success in the metro, and the emergence of standards groups to push interoperability and service improvements, GigE and 10 GigE have built up momentum among carriers. A small but growing number of end users are also clearly on the metro Ethernet bandwagon.

2.7.13   Private Network Adoption of 10 GigE

It will take until around 2006 before 10 GigE is adopted in private (enterprise) networks on a wide scale. There are very few OC-48 enterprise customers today. These enterprises would be the ones to adopt 10 GigE first. Carriers will be the first to deploy it and enterprise end users will be the last, due to issues such as required interfaces and SLAs. Interoperability issues and server and minicomputer interface compatibility will inhibit adoption in private networks for the first few years after 10 GigE is standardized. This means that 10 GigE will gain significant traction in the enterprise around 2004. In most cases, private networks also don't have to worry about stringent QoS demands since they're not forced to perform up to carrier standards and generate revenue.

Key: Look for university campuses and large enterprises to be early 10 GigE adopters.

SLAs will be another area of concern. Service levels should be comparable to leased-line performance, and should cover minimum guaranteed bandwidth, peak burst rate, latency, and restoration timeframes. If customers plan to use GigE as a local loop substitute, the SLA needs to include voice service, which is not Ethernet's strong suit as of 2002. Massive bandwidth cannot necessarily guarantee real-time voice quality. As a matter of fact, a lack of valid, quantifiable SLAs and formal support for VoIP will keep the uptake rate low for GigE-based services in general, according to a Yankee Group study. But this will change; it's just a matter of time and development. Someone will find a way to make voice work well over GigE and 10 GigE.

2.7.14   Beyond 10 Gb

Just as the growth of 10 Mb Ethernet led to the need for 100 Mb Fast Ethernet and as the growth of Fast Ethernet led to the need for GigE, the growth of GigE drove the market to 10 GigE. This trend is not likely to stop anytime soon. Servers—whether they are web servers, file servers, e-commerce servers, or others—must obviously have exponentially greater bandwidth than the customers they serve; otherwise, those customers would feel the effects of inadequate network performance. A prime example is web servers. If the average web browser is using a 56 Kb modem, a server connected to a T1 circuit can simultaneously handle approximately 30 customers. However, the broadband movement has already started—albeit slower than expected—and millions of consumers are now accessing the Internet from DSL and cable networks. These consumers access the Internet at speeds of multimegabits per second, and for them, a service provider limited to a T1 circuit is already unacceptably slow. This requirement to always be faster than the customer by a factor of the number of projected simultaneous accesses drives the need for DS-3 (45 Mb) and greater speeds today, and will drive the need for Fast Ethernet, GigE, and 10 GigE tomorrow.

Key: Engineers are already dreaming of the next step in Ethernet's evolution and arguing over the target speed. 40 Gb speeds (SONET OC-768) have already been demonstrated, so that is possibly Ethernet's next target. However, the Ethernet purists insist that the only logical next step is to move the decimal point one more time to 100 Gbps.

The limits on optical Ethernet bandwidth may just be the limits of fiber-optic bandwidth—perhaps 25 Tbps for the available spectrum on today's fiber, which is still well beyond the capabilities of today's lasers and electronics. Still, extrapolating from current trends gets us to that level in only 5 or 10 years (using Moore's Law).

Standards-based 10 GigE when combined with MANs based on WDM promises to be a viable solution to relieve the metro bottleneck. It offers a hierarchy of speeds, end-to-end protocol consistency, and technical features that are needed by both service providers and end users.

2.7.15   10 GigE Summary

Enterprises are already starting to benefit from the adoption of Ethernet as a WAN technology. In the campus LAN, very large, high-capacity networks will benefit from prestandard 10 Gb links as a tactical upgrade in high-capacity areas. But caution must be exercised to ensure that switch backplanes have the capacity to deliver something approaching 10 Gb link performance. For WAN services, enterprises must negotiate strict SLAs, and for the short term, they should use traditional services as a backup (such as private lines or frame relay).

2.8   Architectures for Metro Area GigE

In terms of network design and topology, service providers are considering deploying metro GigE in either a mesh, ring, or a hybrid combination of these two models.

2.8.1   Mesh Versus Ring

With growth in data driving changes in network traffic, service providers are once again evaluating their network architectures. They're weighing one option against another to ensure the right choice for themselves and their customers. At the core of the debate is the role of ring versus mesh topologies in the emerging metro area architecture.

Through the 1990s, rings were the preferred topology in the access network (local loop). That's not likely to change, even with the advent of new data services. This is partially because access traffic is generally headed to a CO or service POP and little connectivity is required among local sites. In addition, SONET/SDH ring protection techniques are perfect for the geographic limits and traffic patterns typical of the metro environment. Most importantly, the local loop suffers from limited fiber capacity with right of ways that are hard to come by. It's extremely expensive to add another path into or out of a facility. Rings are still the most cost-effective strategy for distributing services in densely populated regions like the local loop.

SONET/SDH rings are still being deployed in today's metro core. However, mesh topologies have proven extremely cost effective for carriers building out new IP/packet services. With the huge growth of data traffic, it makes sense for carriers to consider whether mesh technologies are ready to replace rings in the metro core where traffic patterns are more distributed. For most carriers, it won't be a firm decision. Instead, it's likely they will take advantage of both topologies. The choice is often driven by the services to be delivered and the geography to be served.

Key: It's expected that rings will remain the technology of choice for collecting and aggregating traffic in the access network, where service delivery is the priority. However, in the transport-optimized metro core, rings could give way to a mesh topology. This is because mesh networks ultimately reduce transit time and delay in transport processing (in other words, less hops and table lookups). But before widespread adoption of mesh architectures occurs, mesh-capable provisioning tools must mature.

Ultimately, the access network and metro core must integrate into a carrier-class solution that is resilient, manageable, and integrated with the long-haul infrastructure.

2.8.2   The Mesh Option

A mesh network is defined as a network in which either every node is directly connected to every other node (full mesh) or some subset of nodes (partial mesh). The chief appeal of a mesh is that it provides protection without requiring a 1:1 dedicated protection path. See Figure 2-15 for an illustration of a mesh topology.

A mesh can reduce the cost of the transport network by 30 to 50 percent, depending on level of connectivity, the traffic patterns, and the blueprint that are used. The challenge in mesh networks is that failures affect network efficiency and possibly service levels. It may not be possible to recover all of the circuits, and the time to recovery may not meet the 50 milliseconds required for guaranteed connections. This is not a new issue. In the early days of SONET/SDH, there were regular debates on whether to use a mesh or ring topology in the metro core. The discussions typically centered on provisioning and protection. How does a carrier provision the network efficiently? How should the network be provisioned to guarantee 100 percent protection? How is it possible to dynamically change network protection capacity when new circuits are added? How does the carrier meet the 50 ms

Figure 2-15
Mesh network
topology: all-to-all
connectivity

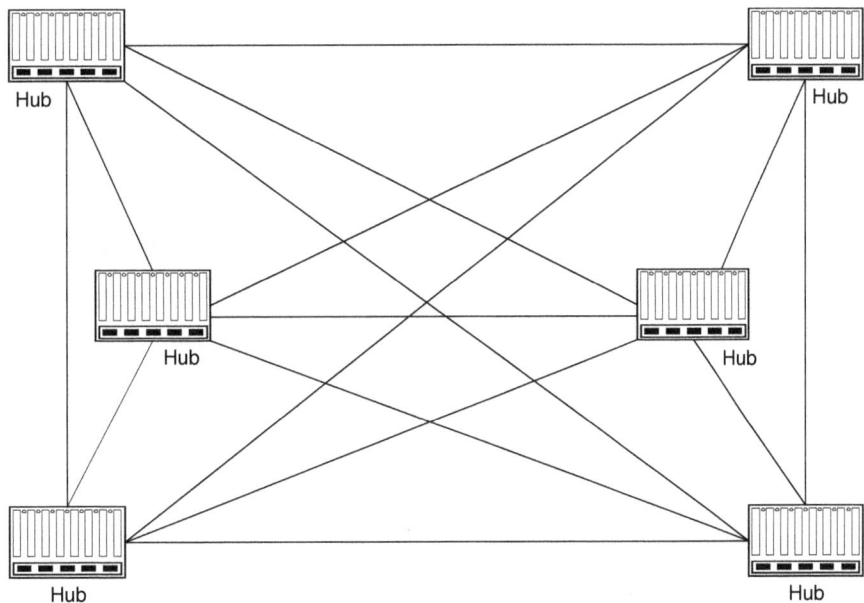

recovery period for key lifeline and SLA-based traffic? For the most part, rings were chosen over mesh, even in the metro core.

Since then, the issues have not really changed, but the technology has changed. Today's faster algorithms have the potential to improve recovery times. In addition, optical cross-connects offer higher bandwidth so ultimately fewer paths are deployed overall, which reduces complexity.

Mesh topologies were originally optimized for IP data networks. They assume distributed connectivity and accept degraded performance in case of failure. Efficiency is achieved by enabling multiple services to share common network protection paths. Additional bandwidth savings can be obtained by using a model that uses alternative paths rather than dedicated protection rings for failure recovery.

A mesh network has no concept of dedicated protection bandwidth. Instead, protection bandwidth must be available to multiple services or circuits to reduce the cost of the network. The degree of protection bandwidth reserved is dictated by the services offered and the algorithm used. For example, a best-effort Internet service may require minimal protection. The trade-off is potentially degraded performance in order to keep costs as low as possible. Conversely, a private-line service requires excess capacity to guarantee service levels, regardless of when or where a failure occurs. This resilience increases overall costs.

There's something else to consider: Mesh-based restoration requires every node to implement the same mechanism. Although ring networks can consist of multiple ring types, mesh recovery standards are not fully developed and don't support this level of interoperability. The *generalized multiprotocol label switching* (GMPLS) recommendation defined by the *Internet Engineering Task Force* (IETF) has emerged as the most likely standard for provisioning end-to-end paths (channels or wavelengths) through a meshed optical network. GMPLS handles not only the logical connections, but it also handles traffic engineering and path recovery (the physical connections).

GMPLS may provide an attractive approach to provisioning mesh networks. Thus far, GMPLS has been aimed at the core of carrier networks, with initial support coming from vendors serving that market. GMPLS may migrate to the metro core, but the true benefit will only be realized when GMPLS is available on an end-to-end basis, including connections to the access network.

2.8.3   The Ring Option and SONET

Key: Rings are still being deployed in the metro core. Although ring technology is usually associated with SONET/SDH, it is also an option for emerging metro Ethernet architectures. This will be the case even more so once the RPR standard is completed in 2003.

Two SONET/SDH standards are governing ring bandwidth utilization and recovery: *Unidirectional Path-Switched Ring* (UPSR) and *Bidirectional Line-Switched Ring* (BLSR). Both are well defined and widely implemented, and provide 50 ms recovery during network failures. However, these standards do not address the bandwidth efficiency, service granularity, and rapid provisioning requirements of packet data applications.

Metro Ethernet rings offer a packet-transport alternative to SONET/SDH. These pure-packet schemes will probably leverage the impending IEEE RPR standard (set for 2003) for bandwidth sharing and recovery. However, even when RPR is ratified, providers will still have to grapple with a variety of issues, including TDM support, QoS capabilities, and end-to-end interoperability with established legacy carrier networks. EoS ring architectures that integrate RPR mechanisms may see implementation before metro area Ethernet using RPR technology.

Key: Equipment costs track the number of physical ports required to terminate fiber connections. A site tied into a ring requires fewer ports than one linked to a large mesh. This is particularly true at a service POP, where fiber from all sites must be centrally terminated.

Another cost factor is at work here. It may be extremely expensive or even impossible to bring several mesh connections to a single site, particularly in dense urban areas. This reality often makes the decision to deploy rings in the local loop a given from a cost perspective.

Clearly, SONET/SDH rings still have the advantage when it comes to provisioning, service management, and billing systems that already exist. Procedures are well understood and proven. As a result, strategies that leverage this significant back-office investment can provide a legitimate cost of ownership advantage to carriers.

2.8.4   A Hybrid Solution

With the technology advances mentioned, many carriers are considering hybrid topologies: partial mesh cores with collector rings to aggregate traffic into the core. New signaling technologies such as GMPLS are also in development, which set up paths dynamically rather than requiring them to be manually provisioned. (GMPLS will be reviewed in Chapter 5.)

Mesh topologies are gaining momentum as next-generation technologies mature (such as optical switches and metro Ethernet core switches). However, in order for mesh architectures to be broadly deployed, three things are required:

- Interoperability standards for path selection and recovery need to be completed.
- QoS mechanisms need to be solid.
- The service provisioning infrastructure needs to be in place.

2.8.5   Architecture Summary

Even with their inefficiencies, SONET/SDH rings are supported by procedures that are fully defined and integrated into existing long-haul infrastructures. Standard recovery mechanisms exist. They're simple and well understood, and offer automatic recovery with full service-level guarantees. SONET/SDH rings also remain the most cost-effective way to distribute services in the local loop. In addition, next-generation technologies that are packet optimized and SONET/SDH capable are becoming available to address the shortcomings of SONET/SDH networks (see Chapter 5).

Simultaneously, there is a lot of momentum behind mesh technologies. They've proven to be very cost effective for data-optimized architectures and are expected to gain ground as data overtakes voice on the public network.

The downside to mesh networks is that they're difficult to provision and have failover times that are unacceptable—at this time—for carrier-class service. However, standards and provisioning tools must and will be developed. Standards such as GMPLS will definitely mature, and faster failover algorithms are already under development.

Key: The bottom line is that the topology decision—mesh or ring—is ultimately driven by fiber cost, the equipment and mix of services already in place, the new services to be offered, and the geographic area to be served.

The good news is that carriers are not forced to overhaul an existing network to accommodate the changes introduced by data. By deploying a packet optimized, SONET/SDH-compliant metro access solution, it is possible to leverage the current network in the migration to next-generation packet services.

2.9   Ethernet in the Access Arena

Along with its foray into the metro arena where it supports enterprise transport, Ethernet is also being seriously considered as a consumer technology for Internet access. It may even mirror DSL deployments at some point in the future, where both voice and data will traverse one Ethernet connection to the nearest *serving wire center* (SWC). Several of these Ethernet-based access technologies are discussed here.

2.9.1   Ethernet in the First Mile (EFM)

The IEEE 802.3ah *Ethernet in the First Mile* (EFM) Task Force was created on November 9, 2001. They are the responsible for developing the topology and PHY layer standards for the use of Ethernet in the subscriber access network. The standards-setting process has a goal of delivering an approved standard by September 2003. Currently, the IEEE 802.3ah EFM Study Group has the following objectives:

- Support subscriber access network topologies.
- Provide a family of PHY layer specifications.
- Support far-end OAM for subscriber access networks.

Access network topologies to be developed include point-to-multipoint connectivity on optical fiber, point-to-point connectivity on optical fiber, and point-to-point connectivity on copper cable.

PHY layer specifications include 1000BaseX (extended temperature range optics), which has a targeted range of 10 km over a single SMF. Also targeted for development is a PHY layer standard for *passive optical networks* (PONs), which is also for distances greater than or equal to 10 km (~6 miles). Gigabit connections over SMF from 1 to 16 km are also under consideration. A PHY for single-pair voice-grade copper up to 2,500 feet with 10 Mbps aggregate speed is included as well.

The schedule for setting an EFM standard is very aggressive and won't likely be met. A Working Group ballot was scheduled for July 2002, and approval of the standard is scheduled for September 2003. Initial trials and demonstrations of prestandard equipment are possible four to six months after baseline approval of the standard.

2.9.1.1   Ethernet Point-to-Point over Fiber (EP2P)   The standard being develop for the *Ethernet Point-to-Point over Fiber* (EP2P) topology is essentially an extension of the current 10000BaseLX spec developed for the IEEE 802.3z GigE standard. The main difference will be transmission over single fiber line using DWDM to split the downstream and upstream traffic into two separate frequencies rather than using a complete fiber pair to separate downstream and upstream traffic.

2.9.1.2   Ethernet Point-to-Multipoint over Fiber   This standard essentially mirrors most of the *Ethernet passive optical network* (EPON) standard that's under development (see the section "Ethernet Passive Optical Networks [EPONs]").

2.9.1.3   Ethernet Point-to-Point over Copper   The objective of this standard is to take advantage of the existing installed base of voice-grade copper cabling without the cost of having to lay new fiber-optic infrastructure. Currently, three separate proposals have been developed to address how this standard should be implemented.

2.9.1.4   100BaseCU   This is a proprietary implementation from Elastic Networks. The company claims a symmetrical throughput speed of up to 100 Mbps up to a distance of ~21,000 feet. This is a *listen, adjust, and then transmit* requirement. Performance with 100BaseCU will depend on the number, type, and distribution of other services in place. Performance will not be exactly the same for every customer at a given loop length; it will depend on the given environment and could change over time. 100BaseCU will not be an always-on service due to potential problem for other interference limited services. Silent periods will allow for channel measurements in order to enhance system performance.

2.9.1.5 EoVDSL This is an open standard that seeks to adapt very high-bit-rate DSL technology for use with Ethernet. Currently, its main developer and cheerleader is Extreme Networks. EoVDSL would use frequency division duplexing technology to eliminate near-end crosstalk. The primary objective is to enable high data rates available over short distances, as follows:

- 52 Mbps downstream/6 Mbps upstream at a range of 1 *kilofoot* (kft)
- 13 Mbps downstream/13 Mbps upstream at a range of 3 kft

2.9.1.6 10BaseT4 This is a more recent EFM proposal known as 10Base T4, where the objective is to enable 10 Mbps full-duplex transmission by transmitting bidirectionally using four pairs of twisted copper wiring, with each pair running 2.5 Mbps to achieve a symmetrical rate of 10 Mbps at a distance of up to 12,000 feet. Loss of one pair of cable degrades the service by 2.5 Mbps, but does not completely halt the service. (Note: Although the 802.3ah Task Force is setting a minimum target of 2,500 feet, no maximum upper limit is being set on the effective range.)

2.9.2 Ethernet's Potential for Broadband Access

By the end of 2001, roughly 62 million households out of the 108 million in the United States (or 58 percent of the total households) will have access to the Internet. Table 2-8 shows a projection of the growth of homes that will have some kind of access to the Internet through 2005. Of those 62 million homes, approximately 18 percent will have migrated to some form of broadband access by 2005.

Table 2-8

Percentage of U.S.
online households
—2000 to 2005

	2000	2001	2002	2003	2004	2005
	52%	58%	63%	67%	69%	70%

Source: Cahners In-Stat Group

Key: There is little doubt that dial-up access remains the primary mode of consumer Internet access in the United States as of 2002. Only the development and widespread adoption of a true killer application will change that, and stimulate a migration of the public at large to broadband technologies. A must-have service needs to be made available. Major regulatory reforms are also necessary to ensure that the competitive playing field is level. This applies in particular to the RBOCs.

With the overlay of Ethernet access standards being developed, the technology has the potential to revolutionize how the residential subscriber relates to the Internet. However, in the near term (2002 to 2006), Ethernet to the residence will be provisioned first over the existing installed copper plant and will eventually move to full EPON systems as fiber outside plant is pushed further out to the customer—into neighborhoods.

Key: Eventually, Ethernet broadband access may be provisioned via full point-to-point fiber access as greenfield and overbuild deployments increasing turn toward fiber as the medium of choice for converged data, voice, and video services. However, this type of provisioning would entail a massive build out of outside plant infrastructure, which would require a substantial amount of capital funding. It may be a bit too soon to presume that a telecom venture of that scope is ready to be undertaken since the industry is still in a major financial funk circa 2002.

2.9.3 Ethernet as Subscriber Access Technology

Given its many advantages, Ethernet will eventually be a key residential subscriber access technology. Companies working to bring this about include Alloptic, Optical Solutions, Salira, and World Wide Packets.

Cahners In-Stat believes that the full benefit of Ethernet broadband access will be derived from its use over fiber media. Rest assured that the RBOCs will work diligently to see how they can optimize their existing copper plant, which cost billions to lay over the last 100+ years.

Technology geared toward the subscriber access network has recently become a hot issue because the subscriber access network has historically been the most widespread bottleneck in the fast and efficient transfer of data, voice, and video from service providers to the residential subscriber and between subscribers as well. Although the public carrier network (sometimes referred to as the *core*) runs at multigigabit optical speeds, and gigabit speeds have become a reality in private LANs with the advent of GigE in 1998, technologies used to connect residential subscribers to the PSTN have suffered from

- High costs
- Limited bandwidth
- Media limitations (copper's inherent restrictions)

Not only does Ethernet broadband access resolve the limitations mentioned for current broadband access alternatives, but the Ethernet protocol also dovetails nicely with the growing tendency for residential subscribers to establish home networks as consumers grow ever more connected.

To date, the primary impetus for the use of Ethernet in the subscriber access network has been on the business side. Companies are trying to reduce costs by simplifying their networks with equipment that's much cheaper than ATM and SONET gear while increasing their bandwidth at the same time. A spillover effect is starting to have an impact on residential subscriber access technologies as equipment vendors and service providers start to realize that the very qualities that make Ethernet a superior choice in the business subscriber access market (its simplicity and reduced costs) also make it an ideal platform to develop for residential subscriber access.

Key: By simplifying access technologies to common platforms capable of delivering inexpensive high bandwidth to both residential and business subscribers, service providers should be able to reap even greater cost reductions through economies of scale and efficiencies in service provision. This could be achieved by using the same aggregation equipment in carrier COs for both market segments—business and consumer.

2.9.3.1 Fiber to the Home (FTTH) *Fiber to the Home* (FTTH) is already in trial deployments in several communities. Several studies have been completed suggesting that building a FTTH network is no more expensive than a full-scale DSL build out or a two-way *cable television* (CATV) upgrade. Estimates show FTTH costing in the neighborhood of $1,000 per home passed. The potential capacity of an FTTH network is much greater than DSL or coaxial cable (CATV infrastructure), so it's clear that for any new build out, laying fiber is the logical choice. Like *Fiber to* *the Business* (FTTB), Ethernet is the least expensive technology providing optical-network access today.

It seems clear that Ethernet services will be coming soon to businesses and to residences soon after. The only questions are how soon, how fast, and how expensive will it be? In many metropolitan areas, some service providers already provide Ethernet services by the megabit of capacity.

Residential applications are more limited and therefore have less bandwidth requirements. 10 Mb Ethernet should be adequate for web surfing into the near future, and the bandwidth needed for voice service is negligible—voice bandwidth is 3.3 KHz. The most bandwidth-hungry application recognized today is TV, namely premium services such as video on demand. A single video channel or movie at broadcast quality requires about 4 Mbps of bandwidth, DVD quality requires only about 9 or 10 Mbps, and even HDTV will probably only require 20 Mbps. These figures are per channel viewed, and the industry should plan on an average capacity of two channels per household at a time. However, even the most aggressive of these numbers imply that a single dedicated Fast Ethernet service to each home is more than adequate for services that exist today.

2.9.3.2 Fiber to the Business (FTTB) FTTB is a reality for most of the Fortune 500 today. A carrier that runs fiber to a business enables the delivery of all of today's communications services and, likely, all of tomorrow's. Ethernet services are the least expensive services that can be provided over that fiber today.

Whether Fast Ethernet is fast enough depends on the business. But the cost of laying FTTB dominates capital expenditures of service providers, and today's Fast Ethernet infrastructure can be readily upgraded tomorrow. The switching cost and Internet-access cost are still much higher for gigabit speeds, but those costs are rapidly declining. As a matter of fact, for Ethernet services between business locations (TLAN service), GigE may already be needed to support file servers, backup servers, and other intranet applications. Most workstations installed today come with built-in Fast Ethernet NICs, implying that the network switching and office-services infrastructure should be significantly faster to avoid bottlenecking.

2.9.4   Securing the First Mile

Even without extensions to the existing 802.3 standard, Ethernet is the logical choice for first-mile connectivity for the following reasons, which we've seen before:

- Predominance in the networking space (LANs, MANs, and CANs).
- A much better cost-to-performance ratio versus other networking technologies.
- The ability to leverage existing technology—a massive embedded base.
- Scalability—10 Mb, 100 Mb, 1,000 Mb (1 Gb), 10 Gb, 40 Gb, or 100 Gb?
- No bridging or protocol translation is necessary.
- Customers love Ethernet because it's simple . . . plug and play.
- EFM completes the end-to-end Ethernet picture (Ethernet everywhere).

Thousands of companies are coming up with ideas to fill the broadband pipe. The nature of consumer usage will eventually drive new applications that haven't been delivered yet, but it won't be long until they're available. Table 2-9 shows how much bandwidth will be required by households that utilize various applications. The total bandwidth requirement per household will be 32 Mbps on average per this illustration.

If Ethernet works as well as currently defined, why enhance the standard for the first mile? Because enhancing the standard may improve and refine the technology specific to this space. Some minimum criteria for success are as follows:

- **Minimum bandwidth requirement** 100 Mbps.
- **Scalability and futureproofing** Avoid infrastructure limitations.
- **Avoid shared media technologies** Home-run connectivity from the head end to the customer with full-duplex links would be optimal.

Table 2-9

Imminent household traffic in megabits

Application	Bandwidth Requirements
HDTV video (one channel)	20 Mbps
SDTV (one channel)	5 Mbps
Web surfing	5 Mbps
Games	2 Mbps
Phone conversation	64 Kbps
Total	**32.064 Mbps**

Enhancement possibilities could include

- A new single-fiber PMD—a PMD transceiver with the following characteristics:
  - Two wavelengths (cost optimized).
  - Bidirectional/full duplex.
  - Media and installation costs could be halved.
- Extensions to layer 2 tagging to increase supported number of VLANs
  - 16 million combination is an ideal number.
  - Increases maximum frame size by four octets.
  - Impacts 802.1p and 802.1q.

2.10 Ethernet Passive Optical Networks (EPONs)

EPONs are an emerging access network technology that provide a low-cost method of deploying optical access lines between a carrier's CO and a customer site. EPONs build on the *International Telecommunications Union* (ITU) standard G.983 for *ATM PONs* (APON) and seek to achieve the dream of a *full-services access network* (FSAN) that delivers converged data, video, and voice over a single optical access system.

PON can support transmission up to 12 miles, depending on the number of customers on the network and the capabilities of the laser signal regeneration, which is why it's only targeted for short metro links. Carriers are just beginning to deploy PON, with players such as Quantum Bridge, Paceon, and Optical Solutions offering systems today.

2.10.1 Evolution of PONs

PONs address the last mile of the communications infrastructure between the service provider's CO, head end, and POP, and business or residential customer locations. Usually, only large enterprises can afford to pay the $3,200 to $4,300 a month it costs to lease a DS-3 circuit (45 Mbps) or optical carrier (OC-n) SONET connection. T1s that cost $375 a month (average) are an option for some medium-size businesses, but most small- and medium-size enterprises and residential customers are left with few options beyond *Plain Old Telephone Service* (POTS) and dial-up Internet access at 56 Kbps. Where available, DSL and cable modems offer a more affordable interim solution, but these services are difficult and time consuming to provision. DSL availability is also constrained based on whether or not the required electronics are in serving COs.

Key: In EPON systems, bandwidth is limited by distance and the quality of existing wiring (similar to xDSL services), and voice services aren't widely implemented over these technologies. The result is a growing gulf between the capacity of metro networks on one side and end-user needs on the other side, with the last-mile bottleneck in between.

PONs aim to break the last-mile bandwidth bottleneck by targeting the sweet spot of bandwidth between T1s and OC-3s that other access network technologies do not adequately address. This bandwidth area also includes a sweet spot that will provide a variety of cost-effective speeds and feeds that don't exist today.

2.10.2 EPONS Versus APONs

The key difference between EPONs and APONs is that in EPONs, data is transmitted in variable-length packets of up to 1,518 bytes (maximum) according to the IEEE 802.3 standard for Ethernet. Conversely, in APONs, data is transmitted in fixed-length 53-byte cells (with 48-byte payload and 5-byte overhead), as specified by the ATM protocol. This format means it's difficult and inefficient for APONs to carry traffic formatted according to the IP. IP calls for data to be segmented into variable-length packets of up to 65,535 bytes maximum. For an APON to carry IP traffic, the packets must be broken into 48-byte segments with a 5-byte header attached to each one. This process is time consuming and complicated. It adds protocol overhead in the form of the notorious ATM cell tax, which adds additional cost to the *optical line terminal* (OLT) and *optical network units* (ONUs). When using ATM in any architecture, 5 bytes of bandwidth are wasted for every 48-byte segment, creating an onerous overhead. By contrast, Ethernet happens to be tailor-made for carrying IP traffic and dramatically reduces overhead compared to ATM. The ideal bandwidth range for EPON systems is shown in Figure 2-16.

Figure 2-16
Ideal bandwidth
range for PONs

RANGE OF OPERATION FOR PASSIVE OPTICAL NETWORKS

SWEET SPOT OF OPERATION

BANDWIDTH (bps)	64K	144K		1.5M	45M	155M	1G	10 G
			DSL					
SERVICES	POTS	ISDN		T-1	T-3	OC-3		OC-192

Ethernet
10BaseT

Fast Ethernet
100BaseT

Gigabit
Ethernet

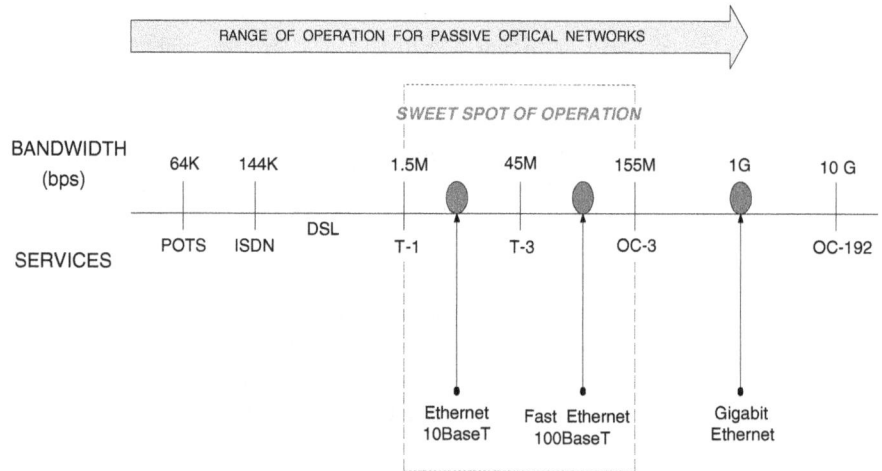

2.10.3   EPON Fundamentals and Benefits

Unlike point-to-point fiber-optic technology, which is optimized for metro and long-haul applications, EPONs are specifically designed to address the unique demands of the access network. Because EPONs are simpler, more efficient, and less expensive than alternative access solutions, EPONs finally make it cost effective for service providers to extend fiber into the last mile. This enables them to reap all the rewards of a very efficient, highly scalable, low-maintenance, end-to-end fiber-optic network.

The key advantage of an EPON is that it enables carriers to eliminate complex and expensive ATM and SONET NEs and dramatically simplify their networks. Traditional telecom networks use a complex, multilayered hierarchical architecture which overlays IP onto ATM, SONET, and WDM. This architecture requires a router network to carry IP traffic, ATM switches to create *virtual circuits* (VCs), ADMs *digital cross-connects* (DCS) to manage SONET rings, and point-to-point DWDM optical links for layer one transport.

A number of limitations are inherent to this legacy architecture:

- It is intensely difficult to provision because each NE in an ATM path must be provisioned for each different service.

- It is optimized for TDM voice (not data) so its fixed bandwidth channels are ultimately inefficient and have difficulty handling bursty data traffic.

- Legacy TDM-based architectures require inefficient and expensive OEO conversion at each network node.

■ They also require installation of all nodes up front (because each node is a regenerator).

■ Older architectures do not scale well because of their connection-oriented VCs.

In the example of a streamlined EPON architecture in Figure 2-17, an ONU replaces the SONET ADM and the router at the CP and an OLT replaces the SONET ADM and ATM switch at the CO. Figure 2-17 illustrates how EPON streamlines service provider architectures when compared to legacy designs and operations.

To summarize, an EPON architecture offers carriers a number of benefits. First, it lowers up-front capital equipment and ongoing operational costs relative to SONET and ATM costs. Second, an EPON is easier to deploy than SONET/ATM because it uses simpler hardware and no outside plant electronics, which reduces the need for experienced technicians. Third, it enables flexible provisioning and rapid service reconfiguration. Fourth, it offers multilayered security, such as VLAN closed user groups and support for VPNs, *IP security* (IPSec), and tunneling. Finally, carriers can increase revenues by exploiting the broad range and flexibility of service offerings available over an EPON architecture. Profit margins will be higher due to lower infrastructure costs. This includes delivering bandwidth in scalable increments from 1 Mbps to 1 Gbps and value-added services such as managed firewalls, voice traffic support, VPNs, and Internet access.

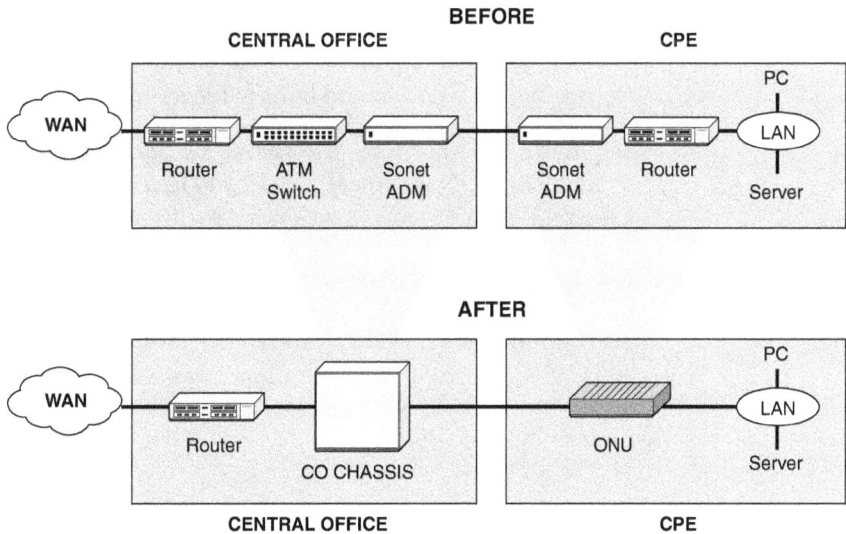

Figure 2-17
Illustration of how EPON simplifies carrier network architectures: before (top) and after (bottom) (Source: iec.org)

The development of EPONs has been spearheaded by several visionary startup companies that feel the APON standard is an inappropriate solution for the local loop because of its lack of video capabilities, insufficient bandwidth, complexity, and higher expense. As the migration to Fast Ethernet, GigE, and now 10 GigE gathers steam, these startups believe that EPONs will eliminate the need for conversion in the WAN LAN/WAN connection between IP and ATM protocols.

EPON vendors are focusing initially on FTTB and FTTC solutions, with the long-term objective being the realization of a full-service FTTH solution for delivering data, video, and voice over a single platform. Although EPONs offer higher bandwidth, lower costs, and broader service capabilities than APON, the architecture is broadly similar and adheres to many G.983 recommendations.

In November 2000, a group of Ethernet equipment vendors kicked off their own standardization effort under the auspices of the IEEE via the formation of the EFM Study Group. Sixty-nine companies have indicated they'll participate in the group, including 3Com, Cisco Networks, Alloptic, and World Wide Packets.

2.10.4   PON Architecture

The passive elements of an EPON are located in the optical distribution network (also known as the outside plant). They include single-mode fiber-optic cable, passive optical splitters/couplers, connectors, and splices. Active NEs, such as the OLT and multiple ONUs, are located at the endpoints of the PON. Optical signals traveling across the PON are either split onto multiple fibers or combined onto a single fiber by optical splitters/couplers, depending on whether the light is traveling up or down the PON.

Key: The PON is typically deployed in a single-fiber, point-to-multipoint, tree-and-branch configuration for residential applications.

The PON may also be deployed in a protected ring architecture for business applications, or in a bus architecture for campus environments and MTU. See Figure 2-18 for an illustration of a PON architecture.

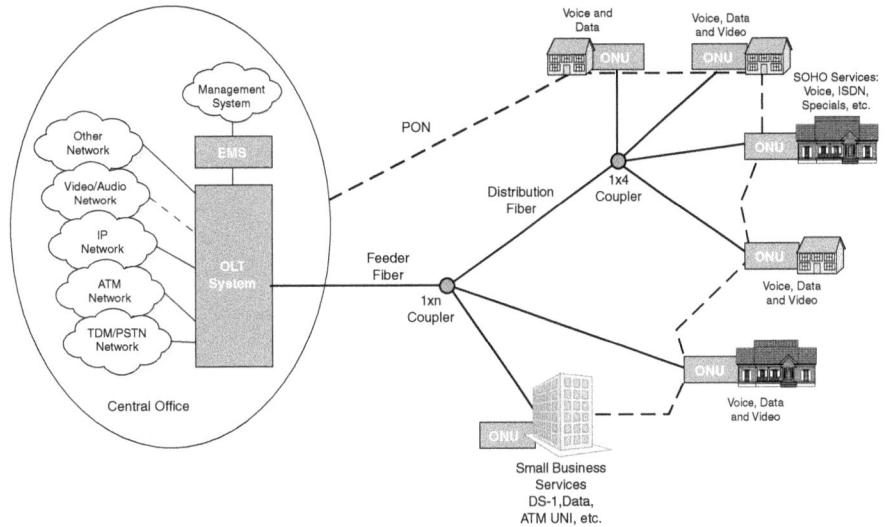

Figure 2-18
Passive and active
NEs of a PON
(Source:
www.iec.org)

2.10.5   Economic Justification for EPONs

The economic case for EPONs is simple: Fiber is the most effective medium for simultaneously transporting data, video, and voice traffic, and it offers virtually unlimited bandwidth. However, the cost of running fiber point-to-point from every customer location all the way to the CO, installing active electronics at both ends of each fiber, and managing all of the fiber connections at the CO is obviously prohibitive.

Key: EPONs address the shortcomings of point-to-point fiber solutions by using a point-to-multipoint architecture in the outside plant portion of the network instead, by eliminating costly active electronic components such as regenerators, amplifiers, and lasers from the outside plant. Active components will always cost much more than passive components. In the PON scenario, the passive component is a type of optical splitter, a prism of sorts. This prism redirects optical signals (lambdas) down different routes in the last miles.

Table 2-10

Comparison of point-to-point fiber access and EPON

Point-to-Point Fiber Access	EPON
Point-to-point architecture.	Point-to-multipoint architecture.
Active electronic components are required at the end of each fiber and in the outside plant.	Eliminates active electronic components such as regenerators and amplifiers from the outside plant, and replaces them with less expensive passive optical couplers that have simpler designs, are easier to maintain, and have longer lives than active components.
Each subscriber requires a separate fiber port in the CO.	Conserves fiber and port space in the CO by passively coupling traffic from up to 64 ONUs onto a single fiber that runs from a neighborhood demarcation point (splice box or manhole) back to the service provider's CO, head end, or POP.
Expensive active electronic components are dedicated to each subscriber.	The cost of expensive active electronic components and lasers in the OLT is shared over many subscribers.

An EPON architecture reduces the number of lasers needed at the CO. Table 2-10 illustrates the key differences between point-to-point fiber transport versus EPON.

2.10.6   Cost-Reduction Opportunities

EPONs offer service providers unparalleled opportunities to reduce the cost of installing, managing, and delivering existing service offerings. For example, EPONs can

- Replace active electronic components with less expensive passive optical couplers that are simpler, easier to maintain, and longer lived.
- Conserve fiber and port space in the CO.
- Share the cost of expensive active electronic components and lasers over many more subscribers.
- Deliver more services per fiber and thereby slash the cost per megabit.
- Enable long-term cost-reduction opportunities based on the high volume and steep price/performance curve of Ethernet components.
- Save the cost of truck rolls because bandwidth allocation can be done remotely. It's estimated that each truck roll costs service providers around $750.

- Use standard Ethernet interfaces, which eliminates the need for additional DSL or cable modems.
- Prevent the need for electronics in the outside plant, which reduces the need for costly powering and right-of-way space.
- Free network planners from trying to forecast the customer's future bandwidth requirement because the system can scale up very easily.

For service providers, all these benefits add up to lower capital costs, reduced capital expenditures, and higher profit margins.

Key: Although the focus here is on cost reductions and related benefits for service providers (carriers), the standards of competition do indeed come into play in this context. If a service provider's costs are reduced, those cost savings will ultimately be passed onto consumers in the form of lower service fees and charges.

In addition to POTS, T1, 10/100 BaseT, and DS-3, EPONs support advanced features like layer 2 and 3 switching, routing, VoIP, IP multicast, VPN 802.1q, bandwidth shaping, and billing. Revenue opportunities from EPONs include

- Support for legacy TDM, ATM, and SONET services
- Delivery of new GigE, Fast Ethernet, IP multicast, and dedicated wavelength services
- Tailoring of services to customer needs with guaranteed SLAs
- Quick response to customer needs with flexible provisioning and rapid service reconfiguration

Now that we've reviewed Ethernet in all its current forms, we'll look at how it is being sold and deployed in the marketplace.

The Metro GigE Marketplace

3.1   Starting with GigE

The GigE players aren't big buyers of optical systems, but they are forming a market for a new generation of Ethernet switches that are *carrier-class* and capable of handling gigabit speeds.

While primarily engaged in building GigE networks, the "pure-play" GigE carriers ("ELECs") approach the market a bit differently. Along with targeting the Fortune 250, these service providers have also tapped into a promising customer base that has only emerged since late 2000: *storage service providers* (SSPs), *Application Service Providers* (ASPs), *Internet Service Providers* (ISPs) and owners of large shared facilities that contain banks of terminating equipment and server farms—the "carrier hotels."

The success of these new service providers depends on their ability to quickly adjust to the changing bandwidth requirements of their own customers. Day-to-day planning, management, and development are required. Deploying too much bandwidth at the wrong time raises costs and overhead for the carriers; deploying too little bandwidth means poor service delivery to their customers. This volatility means these new carriers cannot afford the long provisioning lead times inherent with networks that are based on *Synchronous Optical Network* (SONET).

Driven by demand for high-speed and instantly provisioned broadband capacity in the local loop, one estimate claims that GigE service revenues could reach $657 million by 2003. It's also predicted that *capital expenditures* (capex) will total $1.2 billion by 2003.

Key: Unlike *Digital Subscriber Line* (DSL) providers who require collocation and near-charitable cooperation from the *Incumbent Local Exchange Carriers* (ILECs), the metro GigE players own the networks and operations and can target customers more precisely than the *Data Local Exchange Carriers* (DLECs) of the past.

ILECs are also embracing GigE. One of the most active ILECs in this space right now is Broadwing (grown out of Cincinnati Bell), which has tied GigE to its aggressive strategy to build out intelligent optical networks.

3.2 Challenges and Opportunities for Service Providers

The changing face of the *metropolitan area network* (MAN) landscape offers major opportunities for competitive service providers who are not already locked into legacy infrastructures (such as gear based on *time division multiplexing* [TDM]). Until recently, corporate customers have had to rely on incumbent service providers (*Regional Bell Operating Companies* [RBOCs]) to move data across the MAN by leasing a T1, T3, or OC-*n* service that's provisioned and maintained by the service provider. In most instances, this has involved long lead times for provisioning these fixed circuits plus the extra overhead required to convert Ethernet in the *local area network* (LAN) to MAN transport protocols (such as *Asynchronous Transfer Mode* [ATM], SONET, or frame relay) and back again at the other end of the circuit.

The opportunity now looms for service providers to meet these evolving customer needs by offering new networks with convergence-optimized data services based on GigE.

3.3 The Evolution of the Term "Competitive Local Exchange Carriers" (CLEC)

Since 1999, several new breeds of carriers have emerged. The primary focus of these carriers is to provide transport and/or access in the metropolitan areas of the United States in Tier 1 and 2 cities (NFL cities).

By definition, all of these carriers are known as *Competitive Local Exchange Carriers* (CLECs) because they compete with incumbent telcos. However, due to the nature of their narrowly focused technologies and business models, they are also known by several other new names and acronyms. Even with these new acronyms, they are still CLECs by definition as they compete with the ILECs.

3.3.1   Building Local Exchange Carriers (BLECs)

One of these new terms is *Building Local Exchange Carrier* (BLEC). A BLEC is a competitive metro carrier who focuses their network development, marketing, and sales efforts on targeted buildings in metropolitan areas. These buildings are usually *multitenant units* (MTUs), which are commonly high rises. The BLECs' business premise involves targeting their sales efforts to buildings that are near their own facilities (leased or owned). BLECs focus on signing up enough customers in targeted buildings to justify the expense of running fiber laterals to those buildings from the nearest splice point of their own fiber network in the metro area. The best example of a BLEC is the company Cogent Communications, which will be profiled in this chapter along with many other carriers.

3.3.2   Ethernet Local Exchange Carriers (ELECs)

Another new breed of metro carrier is known as an *Ethernet Local Exchange Carrier* (ELEC). These carriers differentiate themselves by offering suites of service offerings that are entirely Ethernet based. A prime example of an ELEC is the company Yipes Communications, which declared Chapter 11 bankruptcy in April 2002 and reemerged in July 2002. Yipes will also be profiled in this chapter.

The market focus of ELECs is retail, and they target bandwidth-hungry businesses in Tier 1 cities. These businesses are often localized, with multiple branch offices or campuses in a single metropolitan area. They usually have between 20 and 2,000 users, and IT staff is limited. This scenario fits vertical industries such as government, engineering, education, high tech, healthcare, and finance. These groups may need large file transfers, video-conferencing, and *computer-aided design/computer-aided manufacturing* (CAD/CAM) applications.

Another new term springing up in 2002 is *Ethernet service provider* (ESP). For all intents and purposes, this is the same as an ELEC.

3.3.3   Optical Local Exchange Carriers (OLECs)

A breed of metro carrier that offers optical fiber for lease in metropolitan areas also exists. These carriers are wholesalers of optical fiber infrastructure, and they usually sell their services to many other (metro) carriers who use this leased fiber to build their own metro infrastructure. These carriers are known as *Optical Local Exchange Carriers* (OLECs). A perfect example of an OLEC is *Metromedia Fiber Networks* (MFN) (which filed for Chapter 11 bankruptcy in March 2002). MFN's core business is leasing fiber to many other carriers who then use this fiber to build their own networks. It also sells services that run over that leased fiber.

It's important to consider the threat of consolidation during a tight investment market downturn. "Many network service providers will not make it," according to a recent Gartner Group Research Note on Ethernet in the *wide area network* (WAN). "However, the business case is compelling enough that several will make it, while several others will likely be acquired by larger carriers. Further, we expect that traditional carriers will begin to offer Ethernet-based services as well, but at a 25 to 40 percent premium compared to the greenfield players."

3.4   Fiber Availability as a Market Inhibitor

The first fiber-optic communication system was installed by AT&T and GTE in 1977. Since then, optical fiber has steadily overtaken copper cable as the medium of choice for modern-day networks. The performance and reliability advantages are obvious: exponentially higher bandwidth potential and survivability offered by SONET systems. Fiber optics is currently used almost exclusively in the physical layers of WANs around the world, and the development of metro optical networks is well under way.

3.4.1   Why Fiber?

Fiber is the physical media of choice not only because of the increased distance compared to copper cable, but also because of its scalability through the addition of *wavelength division multiplexing* (WDM) capabilities.

Key: A fiber lateral typically costs $50,000 to $100,000 to install, making capital recovery an uncertainty, particularly for such a low-priced service. The leading cost driver is labor, which cannot be mitigated by production volumes. When this restriction is removed, Ethernet becomes much more attractive, as it is in data centers where an increasing number of hosting customers are looking to Ethernet as a low-cost connection to their ISP.

Regardless of demand, the viability of optical Ethernet services are ultimately contingent on the availability of optical fiber to office buildings (MTUs) and other customer locations. If no fiber connection is present, no optical Ethernet services can exist, unless another solution is considered.

Despite the strong demand for broadband connectivity, access connections between corporate LANs and service provider networks continue to be limited because less than 5 percent of the 750,000 U.S. commercial buildings have direct fiber connections to deliver broadband service.

Available connectivity solutions are mostly built on outdated copper-based lines, which cannot offer cost-effective or scaleable broadband access. For example, the current T1 and DS-3 connectivity service can be expensive, difficult to upgrade, and involve long provisioning intervals (30 to 90 days).

Multiple options are available including wireless Ethernet (via *free-space optics* [FSO]), aerial installation on poles, accessing utility rights of way (such as sewer and power), and new underground conduits. However, all of these options, except wireless, are extremely expensive and time consuming.

Key: The bulk of metropolitan fiber is concentrated in core rings connecting COs in the center of dense urban environments, not connecting end customers. This is good news for vendors targeting metro core applications, but bad news for those trying to connect to enterprise customers via optical fiber. It is estimated that only 10 percent of buildings in densely populated urban areas in the United States have direct access to optical fiber.

The fiber capacity situation in the MAN consists of two problems. First, the fiber network runs down major traffic arteries and along elevated train lines—the core portion of the MAN. This part of the metro core is being filled up as fast as it is being lit. Metro *dense wave division multiplexing* (DWDM) could prove useful for expanding bandwidth in existing fiber, but it's still very expensive. Many service providers are using the *net present value* (NPV) analysis to determine what is more cost effective: placing fiber or installing DWDM systems? Despite the effort to lay additional fiber strands, new deployment cannot keep up with the increase in traffic.

The second problem is even larger. These intracity legs of fiber are not being deployed widely enough in the downtown areas of Tier 1 markets, much less their suburbs or in smaller Tier 2 or 3 markets. There also are not enough customer premises being linked with last-feet connections—the several hundred yards of reach that connects many buildings to metro access or core rings. Ultimately, fiber laterals need to be widely installed to enable more customers to be connected, but the economics don't seem to merit digging up streets in all places at this time. The initial areas being targeted are not even fully constructed. A huge endeavor such as this takes time and significant investment. Depending on the city in question, it can cost anywhere from $200,000 to $1,000,000 per mile to lay fiber in a metro area (including construction labor, material, and permits). Technology such as that being employed by service providers like CityNet Telecommunications—blowing fiber through sewer tunnels—is promising. Yet simply envisioning the work it takes to fully wire both the MAN core and the last-feet access bridge is daunting by itself.

Many analysts see DSL and cable modems as just stopgap access measures until fiber can be run directly to the customer premise. Many observers actually think that fiber will be necessary to fully enable a future involving entertainment content on demand for residential customers and applications such as videoconferencing for businesses. The lack of last-mile fiber is likely forestalling development of the very applications that are needed to mitigate the long-haul fiber glut—to fill the dormant capacity in intercity links.

Key: All the recent talk of a fiber glut deals specifically with the long-haul portion of the network, not the MAN. This is mainly because these intercity, long-haul networks have had more time to be built and now have more carriers to build them (such as MCI WorldCom, AT&T, Sprint, Level 3, Qwest, and Global Crossing). One reason for this is because competition has existed in this arena for almost 20 years, whereas real competition has only existed in the metro since the 1990s.

The onset of DWDM and its deployment in the long-haul backbones of the *interexchange carriers* (IXCs) have also contributed to this glut situation.

Another difficulty that is not often mentioned is dealing with local governments. It's a big problem for service providers, according to Jason Knowles, a current analysis analyst. "Municipalities are gouging them, charging them to repave entire streets just for digging a trench. They may be better off partnering with other carriers [for last-mile access]."

3.4.2   Expansion of Raw Fiber Capacity

A fact of life in the MAN marketplace has been a significant proliferation of raw fiber capacity, which can also be called *dark fiber*. The availability of underutilized dark fiber is steadily increasing as municipalities, utilities, and traditional telephone service providers routinely lay more fiber capacity than they can actually put into service. With the cost of the fiber itself representing a relatively small part of the total cost of putting the fiber in the ground, it is typically much cheaper to lay more fiber pairs in a single pass than to risk having to add more later. This is consistent with the previous statement that 75 percent of cable construction cost is due to labor charges, even inside of buildings.

In addition to traditional telephone companies, many other entities that already own metropolitan rights of way have also chosen to invest in the installation of fiber cabling. For instance, municipalities and utilities have laid fiber along their existing right-of-way corridors to connect their own locations together, and they lease excess capacity to service providers and other corporate entities.

3.5   Metro Area Ethernet Requirements: The Carrier Perspective

Can a LAN technology meet carrier requirements? Eventually, yes. As Ethernet technology moves into the MAN and WAN, skeptics have been quick to attack Ethernet switches/routers by describing them as unreliable, lacking carrier WAN interfaces, unable to implement stable and robust routing protocols, and unable to deliver the carrier standard of *five 9s* of stable uptime (99.999 percent availability, which equates to 5.3 minutes of downtime per year).

Entrenched carriers and service providers want to preserve their massive investment in legacy infrastructure. As they see their market share erode by competition, however, they will want to keep pace with the competition by offering a variety of applications and new services. Incumbent carriers will invest in new hardware platforms that can be deployed over their current (cable) infrastructure and will phase in the equipment on an as-needed basis, unless the business case dictates otherwise.

As more local fiber loops are rolled out to existing customers who require higher bandwidth and to newly constructed developments, the number of service delivery platforms the providers have will increase, as will the choices for local access. Various flavors of Ethernet will be seen as tools in a service provider's toolkit of provisioning options.

Although service providers and end users would like to make the transition to a simple, low-cost access strategy in the metropolitan network, the reality is that this will take time. Ethernet has been embraced as a simple, low-cost solution for achieving bandwidth greater than the T1 access speed that most enterprises use today. However, can a LAN technology displace the vast investment the incumbent service providers have in SONET infrastructure? Can the emerging service providers last long enough to provide viable competition on an all-Ethernet platform? These are key questions to ask and the answers will shape this marketplace for several years to come.

3.5.1   Carrier Value Proposition: The Value of Optical Platforms

Carriers are realizing new revenues by cost effectively introducing high-speed, protocol-transparent wavelength services via new low-initial-cost platforms. For example, as of 2002, DWDM systems are available at a cost of about $25,000 per wavelength (both sides included). In fact, many carriers are realizing the full return on these investments in less than one year because the carriers and the enterprises recognize the new relationships for what they clearly are: win-win scenarios. The affordability of the networking solutions that enable WDM services to be used enables carriers to target market segments they could not previously support. Because the cost for the carriers is relatively low, they are pricing these services aggressively. Because the cost for the enterprises is low, demand is increasing significantly. Carriers can offer *enterprise system connectivity* (ESCON) services that are less expensive than standard DS-3 service and can provide GigE, Fibre Channel, and other protocol transport at nearly the same price.

Until just a few years ago, incumbent carriers were reluctantly willing to sell dark fiber to enterprise customers who dealt with the networking issues themselves. Today, though, many carriers have stopped all efforts to sell dark fiber. Why? Because selling a service means winning a customer; selling dark fiber may mean feeding a competitor (a seller of dark fiber). The primary benefit to carriers of selling optical services is twofold:

- *Carriers create new revenue streams*. From a single, flexible platform, carriers can lease protocol-transparent, high-speed LAN, and *storage area network* (SAN) services to Fortune 500 organizations, banks, governments, and financial institutions requiring up to hundreds of high-speed connections as well as small and medium-sized enterprises requiring Ethernet services.

- *Carriers futureproof their networks*. The deployment of a sophisticated optical-networking platform lays the foundation for the all-optical network. A complete optical-network infrastructure will include

DWDM systems, optical gateways, and optical cross-connects. It will filter through the entire network (the enterprise access, metropolitan, regional, and long-haul backbone segments). A carrier that implements an optical-networking service platform today positions itself to be at the forefront of this developing trend.

3.7   The Metro Ethernet Forum (MEF)

In May 2001, several networking companies set up a group to promote the use of optical Ethernet technology in metropolitan networks. Calling itself the *Metro Ethernet Forum* (MEF), the group is comprised of 37 ILECs, and other service providers and networking companies. The MEF met for the first time at the SuperComm telecommunications convention in June 2001. Rapid increases in the use and interest in GigE for metro networks led to the creation of the group. It helped that GigE was the most written-about technology in first half of 2001. The MEF's mission is to enhance public awareness of GigE through advocacy at trade shows and a public web site. The group's leaders also want to persuade existing industry technical groups to address metro Ethernet issues as quickly as possible when they arise.

The group will only initiate technical work when other groups will not—when a development void forms. The MEF doesn't intend to compete with existing standards organizations.

In August 2002, the MEF said it had made significant progress in defining specifications for employing Ethernet as a transport infrastructure and service offering in MANs. At their quarterly meeting in Montreal, attended by more than half the forum's 70 member companies, the organization's Technical Committee reached consensus on 16 technical documents and moved them to "draft status" of beyond. The documents pertain to the following areas: Ethernet services, protocol and transport, management and architecture. The next step, according to forum President Nan Chen, is for member companies to develop products that comply with these specifications and test them for interoperability. The Forum plans to host a public interoperability demonstration in the early summer of 2003 to coincide with a major trade show or conference, according to Chen. Three incumbent carriers—SBC, BellSouth, and France Telecom—have representatives on the board of directors of the MEF. The carriers are driving rapid agreement among Forum members on technical specifications, according to Chen's claims, which may be pivotal toward establishing the success of Ethernet in the MAN—and the Forum—because they'll be the ones buying Forum-compliant products.

3.8   Market Inhibitors

The following issues could inhibit the metro Ethernet market's capability to take off as it should:

- Scarcity of optical fiber in metropolitan access networks
- Availability of alternative technologies (such as free-space wireless and private fiber)
- Bandwidth overkill
- An unstable CLEC environment (a reality in 2002)
- ILEC inertia
- Potential roadblocks in standardization efforts, particularly *Resilient Packet Ring* (RPR)

3.9   Service Provider Segments

This section reviews and explains different types of service providers in the metro Ethernet marketplace, their marketing approach, and details on their service offerings.

3.9.1   FSO Laser Strategy (Free-Space Wireless)

FSO is a fiberless, laser-driven technology that supports high bandwidth, with easy-to-install connections for the last-mile and campus environments. It has been in use by the U.S. military for a number of years primarily in naval ship-to-ship communications. FSO systems are starting to gain acceptance in the private marketplace as a solution to replace expensive fiber-optic-based solutions. FSO is also called *wireless laser* or *wireless optical*. These platforms transmit light between two points, often from rooftop to rooftop, at speeds from 10 Mbps to 2.5 Gbps. The main advantage of FSO is speedy deployment. Vendors tout new customer installation lead time to be within a few days. This is in stark contrast to waiting weeks or months to add more capacity to a local fiber network. Disadvantages of FSO include lower reliability due to line-of-sight issues (such as fog, rain fade, snow, birds, and so on). This provider segment focuses their marketing and network deployment efforts on selling the FSO alternative. In other words, places where fiber-optic cable doesn't exist are prime areas for these carriers to target their services to enterprise customers. The key drawback with this segment is that FSO is still inhibited by distance limitations. Hopefully, FSO technology will not suffer the same fate as other rooftop wireless technologies such as *Local Multipoint Distribution Service* (LMDS), the technology that was the foundation of the now-defunct CLEC Teligent. Several of the major providers in this space will be profiled in this chapter.

3.9.2   Wholesalers and Dark-Fiber Providers

This section reviews companies who sell dark fiber and other carriers who wholesale metro Ethernet to retail-based service providers. Dark fiber providers are becoming a significant competitive threat to metro Ethernet service providers.

3.9.2.1   American Fiber Systems, Inc. (AFS)   AFS was launched in August 1999, receiving first-round funding from Sierra Venture Partners and Lucent Venture Partners in April 2000. In November 2000, AFS obtained its first customer. By January 2001, second-round funding was completed. AFS is currently building networks in six cities throughout the United States.

AFS designs and deploys high-capacity dark-fiber networks in the metro areas of Tier 2 and 3 American cities. Its customers can lease the company's fiber networks on the basis of long-term infeasible rights of use. AFS markets its services to carriers and other service providers.

As former executives of some of America's most successful communications companies, the founders of AFS saw a growing need for new technological solutions to the Internet access in smaller metro areas of the United States. With bandwidth demand nearly quadrupling every year, copper-cable and legacy fiber-optic networks are constantly jammed with data. This phenomenon is not restricted to major metropolitan areas; it's also occurring in business districts of mid-sized cities all across the United States.

AFS is a dark-fiber metropolitan infrastructure provider dedicated to enabling aggressive companies to profit from the burgeoning communications market without waiting years to do it. They design, build, lease, and maintain high-capacity, high-bandwidth dark fiber-optic networks, which are completely connected to a city's most important telecom *points of presence* (POPs), such as

- ILEC and CLEC COs
- ISP and ASP facilities
- Interexchange carrier hotels
- Wireless provider and cable company head ends
- Fortune 1000 companies

AFS' FreedomRing™ networks have built-in redundancy for total reliability. Fiber can be leased exactly at the capacity needed, on a strand-by-strand basis, and at a fixed price.

AFS also offers a wide array of professional services designed to help customers light, operate, monitor, and maintain their dark-fiber networks. AFS has a dark-fiber infrastructure all over the United States, and detailed route maps are available at their web site. AFS has completed dark-fiber networks in Kansas City, Minneapolis/St. Paul, Nashville, Cleveland, and Salt Lake City.

Key: With only 2 strands of fiber, the entire contents of the 18 million books in the Library of Congress can be transmitted from Washington, D.C., to Fresno, California, in less than 4 minutes!

3.9.2.2 Cambrian Communications, LLC Cambrian Communications, based in Fairfax, Virginia, is a wholesale MAN provider of broadband services in the central business districts of Washington, D.C., Baltimore, Philadelphia, New York, and surrounding high-growth suburban markets. Cambrian's suite of bandwidth services includes GigE, 10 GigE, and optical wavelength (lambda service), private line, collocation, and peering connection services. Cambrian is targeting carriers and enterprise customers located in suburban markets that are within 15 miles of its intercity network.

As a Cisco-powered network provider with more than 24,000 miles of fiber, Cambrian offers scalable services that give businesses flexibility and control. Cambrian is a facilities-based service provider, delivering end-to-end, all-optical network solutions to meet the needs of carriers, emerging service providers, and major private network owners.

Chapter 3

In May 2002, at a time when many telecommunications companies were struggling just to survive, Cambrian Communications launched its high-speed IP+Optical network in and around Washington, D.C., Baltimore, Philadelphia, New York, and surrounding high-growth suburban markets.

Cambrian's high-speed Cisco-powered IP+Optical network provides diverse, redundant regional back haul and metro access through a backbone network that connects five initial MANs. Carriers and service providers can connect to commercial buildings and/or carrier POPs at speeds from DS-3 to 10 Gbps within a metro market or between metro markets.

A key reason for Cambrian's success is its unique market niche—providing broadband connectivity in underserved high-growth markets in and around the data-intensive Washington, D.C. to New York corridor—an area where an estimated 40 percent of all U.S. telecommunications traffic and 60 percent of European traffic terminates. In these areas, Cambrian provides wholesale broadband services to carriers and services providers seeking a low-cost alternative to building their own networks or using the legacy networks of ILECs. Figure 3-1 is a map of Cambrian's network.

Figure 3-1
Cambrian Communications network map: northeast United States

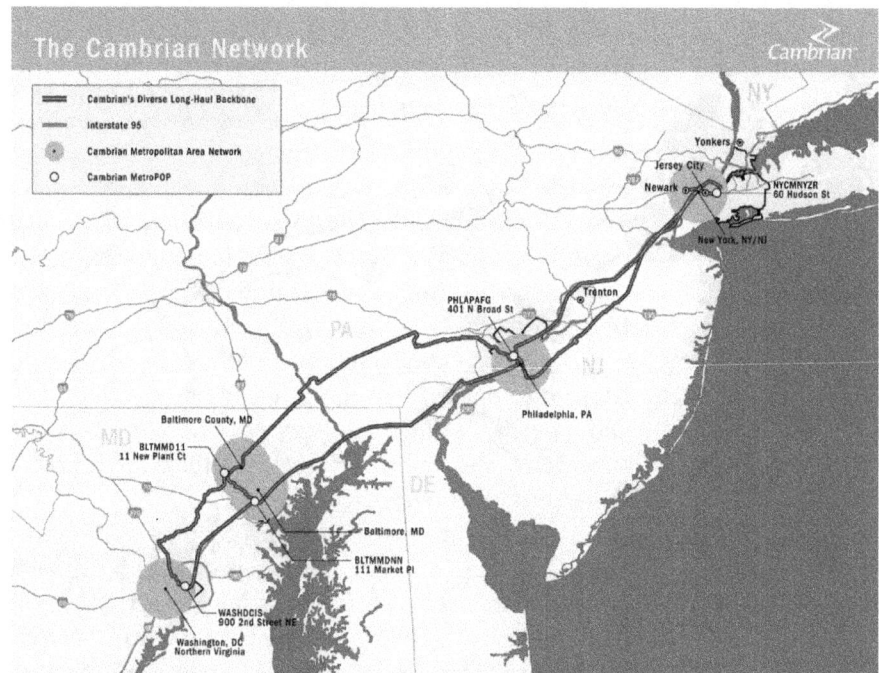

3.9.2.3 Level 3 Communications, Inc. Level 3 is sometimes referred to as a *next-generation IXC*. The company is building an international communications network optimized for the *Internet Protocol* (IP). Its customers include IP-intensive companies whose business operations and services are Internet related. It's aiming to become dominant in managed bandwidth, providing *enabling* services to wholesale customers. The company is using softswitches (standard commercial servers functioning as circuit switches) for managed services and, most recently, long distance.

In 30 months, Level 3 built a 20,000-mile multiconduit intercity network and 33 multiconduit metropolitan networks in North America and Europe. Additionally, the company constructed a transatlantic cable system connecting North America and Europe, and has secured 5.6 million square feet of technical space in 74 data centers serving 57 North American markets and 9 European markets.

On the wholesale (dark-fiber) side, Level 3 offers a product called *(3)Link Dark Fiber (Intercity and Metro)*. This service gives carriers and service providers the infrastructure required to own a fiber-optic network without the burden of constructing a network themselves. (3)Link Dark Fiber service includes optical fiber cable, collocation service, and services where Level 3 will provide line facility space, power, operation, and maintenance of the network as well as other enhanced services.

The intercity features of (3)Link Dark Fiber include

- Approximately 16,000 intercity route miles in North America connecting more than 150 cities
- 3,600-mile Pan-European network
- High fiber counts in Level-3-owned conduits, utilizing the latest generation of optical fiber technology
- State-of-the-art *running-line* collocation facilities

The U.S. metro features of (3)Link Dark Fiber include

- Approximately 1,800 route miles of upgradeable metro fiber networks worldwide, and growing.
- 26 metro markets in North America with more than 130 loops.
- Access to more than 400 strategic on-net buildings.
- Availability of leases with under 20-year *indefeasible right to use* (IRU). This is important because many wholesale providers of dark fiber require 20-plus-year leases. This high cost can sometimes mean the difference between staying in business or declaring Chapter 11 bankruptcy for some carriers.

Level 3 complements their intercity fiber with an extensive metro fiber network. The result is a true end-to-end infrastructure solution.

On the facilities side, Level 3 is a major player in the expanding market for collocation services. The company is building more collocation space worldwide than any other telecommunications company. Collocation is important from the metro Ethernet perspective because it delivers the means for upstart service providers to not only lease fiber from Level 3, but also to use these collocation facilities to house their equipment if desired. It enables startups to do one-stop shopping. (3)Center Collocation facilities include

- State-of-the-art facilities typically sized between 20,000 to 80,000 square feet
- 24×7 access with palm scan security and closed-circuit video surveillance
- Locked cabinet space
- Uninterruptible power supplies with eight hours of backup

Figure 3-2 illustrates Level 3's global network backbone.

Level 3 was impacted by the industry meltdown in 2002, like all other telecom carriers. However, somehow they managed to survive without going

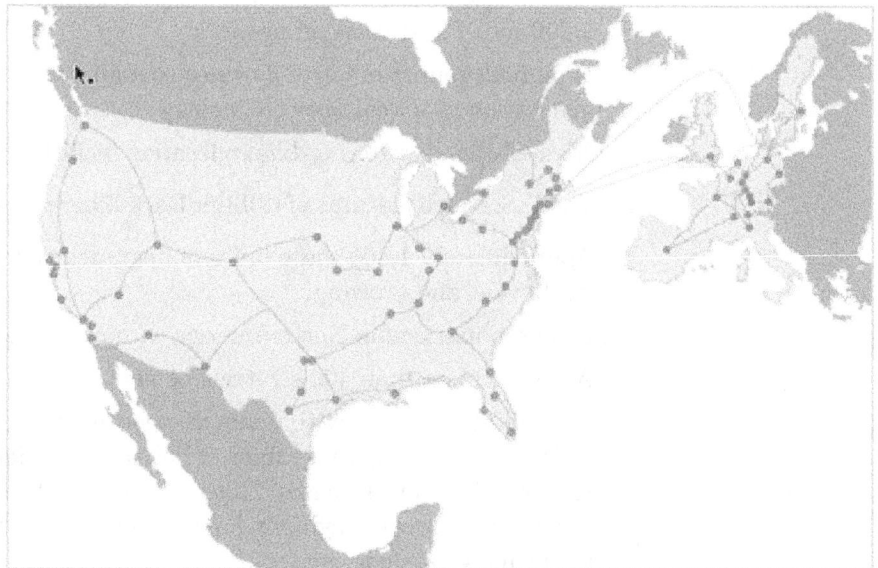

Figure 3-2
Level 3's global network

into Chapter 11 bankruptcy. As a matter of fact, in July 2002, Warren Buffet supported the company with $500 million in capital funding. Warren Buffet is the Chairman and founder of Berkshire Hathaway, a major, well-known capital investment firm. He is widely known as the world's "greatest stock market investor" and holds assets of over $36 billion.

3.9.2.4 Metromedia Fiber Networks, Inc. (MFN)

MFN was founded in 1993 and is a leading provider of digital communications infrastructure (dark fiber). The company combines an extensive metropolitan area fiber network with a global optical IP network and managed services to deliver fully integrated, outsourced communications solutions for Global 2000 companies. The all-fiber infrastructure enables MFN customers to share vast amounts of information internally and externally over private networks and a global IP backbone, creating collaborative businesses that communicate at the speed of light.

MFN is changing the way that bandwidth is priced. When dark fiber is leased from MFN, the cost of the fiber will never go up, no matter how much bandwidth is utilized. MFN fiber is priced at a fixed cost.

> *Key:* MFN offers unmetered, unshared, and unlimited bandwidth via dark fiber at a fixed cost.

With the currently available technology, a single strand of fiber can transmit over 2 *terabits* (Tb) of information per second, and this number is constantly increasing. As the only user of the fiber you're leasing, a company can be assured that no other traffic is on their network and no one can access proprietary data. In other words, it's not just secure, it's *sovereign*.

3.9.2.4.1 The MFN Network

MFN is breaking ground in new places all the time, adding fiber miles and passing more buildings and COs every month. Their expansion plan will consist of approximately 3.6 million fiber miles across 67 cities in North America and Europe.

MFN focuses on deploying fiber infrastructure at the MAN level within larger markets and at the WAN level between those larger markets. The company primarily concentrates on selling and leasing its fiber, but will also assemble and manage DWDM-lit networks for its customers. MFN is a key partner for GigE providers, providing them dark fiber across the MAN to the customer premises. Some of these customers are Yipes, Telseon, and Cogent. Figure 3-3 shows a map that shows the cities where MFN is operational.

Figure 3-3
MFN American fiber-
optic footprint
(Source: mfn.com)

MFN Metropolitan Fiber

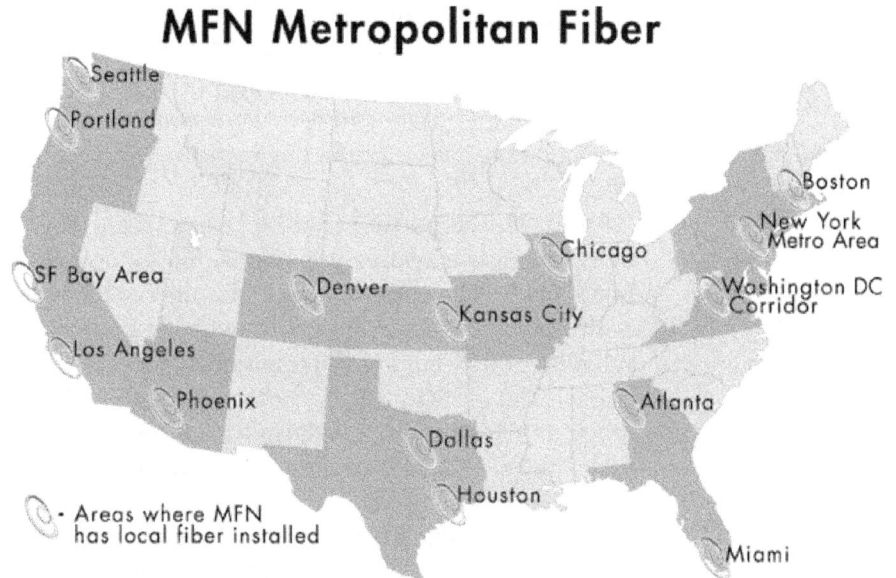

3.9.2.5 OnFiber Communications, Inc OnFiber is a wholesaler of fiber-optic resources. OnFiber Networks acquired the assets of another major dark-fiber network operator, Sphera Optical Networks, in May 2002. OnFiber Communications, Inc. is constructing and operating fiber-optic local access networks in high-growth business corridors in major U.S. metropolitan areas. Like so many others, OnFiber is trying to solve today's greatest communications challenge—the last-mile bandwidth bottleneck in metro areas. By providing end-to-end optical connectivity between service providers (such as carriers, ISPs, ASPs, and SSPs) and their business subscribers, OnFiber's networks enable service providers to take advantage of a wider variety of high-demand broadband applications. The exploding demand for broadband access and lack of suitable connectivity solutions for businesses have created a costly last-mile bandwidth bottleneck for service providers and their business subscribers. OnFiber is approaching this problem by constructing fiber-based local access grids that connect long-haul networks to end users.

The main services offered by OnFiber are as follows:

■ **Local building access** OnFiber provides fiber-based local access circuits that connect business subscribers with their service providers. These local tail circuits can be used to terminate long-haul and Internet traffic or for connecting private lines to other buildings within an OnFiber local access grid.

Figure 3-4
OnFiber's nationwide
MAN rings (Source:
spheranetworks.com)

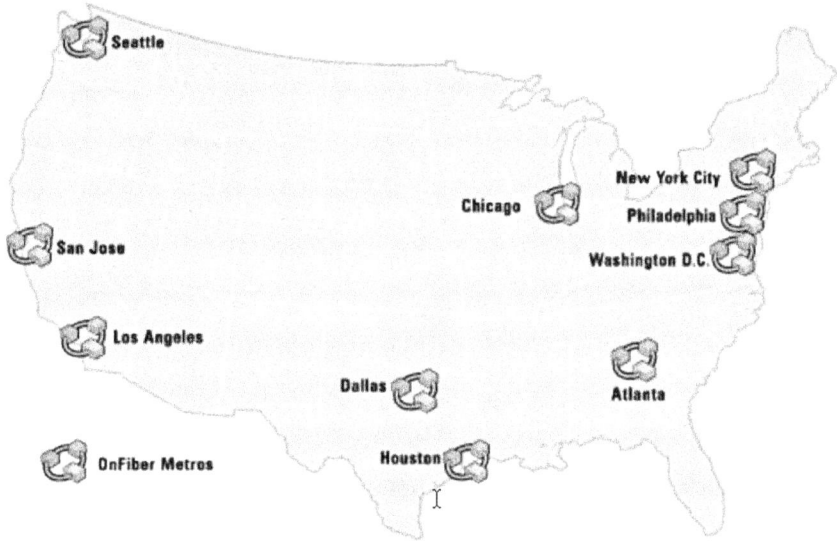

Figure 3-4
OnFiber's nationwide MAN rings (Source: spheranetworks.com)

■ **Metro transport** OnFiber offers high-bandwidth circuits for interconnecting a customer's metropolitan POPs or multiple buildings within a metropolitan area. The primary customers for these services are service providers and large businesses requiring connectivity within a metropolitan area.

Figure 3-4 depicts cities where OnFiber has operational MAN fiber rings.

3.10 Transport Versus Access

To make sense of the competitive landscape, Table 3-1 shows the distinction between two fundamentally different metro business models: transport and access.

There are two categories by which communication systems work in the metro portion of the public network: transport and access.

■ Transport models focus primarily on two data-intensive segments: data centers and carrier interconnection. This market involves highly aggregated data traffic (OC-3 and higher) that originates and terminates in fewer than 50 sites within a metro. Customer acquisition costs and construction of lateral access facilities are minimal.

Table 3-1

Transport versus access

	Transport Segment	Enterprise/Access Segment
Products	• Connections between data centers • Connections between aggregation points (such as POP to POP)	• Internet access • Private line/special access lines • *Transparent LAN* (TLAN) • Voice service
Network location	• Up to 50 data centers, carrier hotels, COs, and POPs	• Hundreds of enterprise buildings per metro (MTUs, *multidwelling units* (MDUs), and single-tenant buildings)
Current traffic characteristics	• Large data flows (OC-3s or greater) • High utilization circuits • Limited number of customers per metro • Increasingly intrametro	• Smaller data flows (DS-0s to DS-3s) • Multiple protocols (ATM, Frame Relay, OC-*n*, and *Transmission Control Protocol / IP* [TCP/IP]) • Low utilization per line • Large number of customers spread out geographically within a metro
Customer needs	• Carrier class—99.999% reliability • Enforceable *service level agreements* (SLAs) • Fast—but not necessarily instant—provisioning of large pipes	• Broad reach • Low price • Instant provisioning and upgrading • Low customer acquisition cost

Source: McKinsey and Co., Lehman Brothers

▨ Access models focus primarily on the data-intensive, large-enterprise, and MTU segment. Data traffic for Internet access and private data links typically requires less than DS-3 bandwidth in the current market environment.

Key: In a typical Tier 1 metro (NFL cities), more than 70 percent of traffic is generated by less than 300 of the 10,000+ MTUs.

Customer acquisition and last-mile connectivity feature prominently in the business models of access players.

3.11   Assessment of the Metro Players

This section contains an assessment of service providers in the metro space. Each has its own approach and characteristics.

3.11.1   ILECs

With the current market opportunity, the emergence of new optical technologies, their access to conduits, and their interest in defending their local franchises, the ILECs would be expected to respond eagerly to the explosion of demand for bandwidth.

Key: The most likely scenario is one where the ILECs launch Ethernet-based products and build these networks on an as-needed basis in parallel with legacy network expansions (where necessary). Over time, as newer technologies become cost effective (such as *multiservice provisioning platforms* [MSPPs], which are reviewed later in this chapter) and the ILECs overcome internal resistance to major network infrastructure changes, networks based on entirely new network technologies and platforms will slowly be built out. These platforms will allow for the integration of legacy network protocols (such as SONET and ATM) and newer protocols (such as Ethernet and xWDM) to peacefully coexist in one network box.

If metro Ethernet-based services are offered by ILECs, they're usually sold on a sell-and-deploy basis.

3.11.2   Dark-Fiber Companies

A number of players are installing substantial amounts of dark fiber within metropolitan areas, positioning themselves as leading providers of this key raw material. MFN is the single largest dark-fiber player in the United States.

3.11.3 Turn-Key Solutions (Self-Built Networks)

Out of frustration with traditional network service providers, some larger customers such as AOL have built their own data networks. These entities have leased dark fiber, purchased equipment, and lit their own networks. Because they lack the aggregated local traffic to achieve economies of scale from a transport perspective, it's most likely that these companies have little long-term interest in owning and operating data networks. Still, they've pursued this avenue to satisfy their immediate requirements.

3.11.4 IXC Approach

Backbone long-haul network providers (IXCs) have built metro transport rings to connect their hubs with other carriers, directly, and at carrier hotels. At the same time, they have aggressively marketed to the high-bandwidth data centers that they covet as customers.

3.11.5 New Transport Entrants

The economics of metro transport networks suggest that individual Tier 1 markets can support only two or three nonincumbent players; therefore, an unsustainable competitive structure has been in place as of early 2002. This structure could eventually lead to aggressive pricing pressure and developments such as partnerships and even consolidation. This activity has already begun. The ELEC Yipes Communications (which filed for Chapter 11 bankruptcy in May 2002 and reemerged July 2002) purchased the assets and building rights of the defunct broadband office in July 2001. In October 2001, Yipes also acquired the customers of Allied Riser Communications, another BLEC that folded. MFN and XO Communications both declared Chapter 11 bankruptcy in the spring of 2002. Their futures are uncertain, but don't count them out yet.

Looking Glass Networks and Sphera Optical Networks start with a greenfield to build and operate a metro network unhindered by legacy protocols or designs. The key to the success of these players will be their ability to rapidly establish a broad presence in the top four to five Tier 1 metros.

This presence could be initially achieved by rapidly securing main anchor customers followed by regional consolidation to help reach scale and secure competitive positions.

Key: Compatible *operational support systems* (OSSs) and complementary network architecture strategies should be a prerequisite to implementing regional consolidations of carriers.

With their existing customer relationships, extensive fiber networks, and local scale and scope advantages, the large ILECs are well positioned in the metro market. However, they are taking measured steps to capitalize on the metro opportunity.

With their intercity backbone networks and high-bandwidth data traffic, the major IXCs also have natural advantages and interest in the metro transport space. Metro transport networks are logical extensions of national backbone infrastructures. Accordingly, IXCs such as WorldCom and Level 3 are building transport rings in most Tier 1 metros using metro DWDM and SONET-lite architectures that are compatible with their backbone technologies.

Key: The competitive battle within each metro area features the ILECs defending their natural franchise, the IXCs extending their networks to high-traffic data centers, and the new players scrambling to establish themselves at a moment in time when demand still exceeds capacity by a wide margin.

ILECs, *Competitive Access Providers* (CAPs), and CLECs have traditionally served enterprises. Since 2000, the industry has seen the entrance of GigE-powered new entrants into the marketplace that are using disruptive pricing in an attempt to carve out a market position (such as Cogent).

Network access has always been a rough business. The natural advantages of incumbents and the aggressive business strategies of well-capitalized new entrants using the latest technology have made profitable entry difficult. Lessons from DLECs, CLECs, and CAPs suggest that the new GigE players will also find securing a profitable niche a challenging proposition.

3.12   Success Factors for New Entrants

The success of new entrants will largely depend on their ability to overcome high costs, niche product offerings, and long sales cycles for target customer segments. As CLECs have found, the strong existing customer relationships, entrenched asset base, and bundled offerings of incumbents (such as voice, data, management, and so on) present a difficult competitive reality for those positioned with even the best technologies.

As a result, the current incumbents with the broadest customer relationships and lower-cost structure may be best positioned to deliver the GigE solution. The ILECs' obvious interest in defending captive voice revenues from major enterprises creates another complication. Although voice revenues are stagnant, they still constitute the majority of ILEC revenue in the metro market. Competitors who want to connect high-bandwidth pipes to major enterprises may initially pursue incremental data traffic, but will ultimately search for ways to siphon off voice revenues as well. The ILECs are likely to implement major defensive strategies in this scenario.

New GigE players such as Cogent and Telseon face major impediments and will need to efficiently evolve their business models to ensure profitability. To improve their chances of success, their strategies should focus on the following select strategic and operational initiatives:

- **New entrants**   Consider lowering or eliminating peering costs through a merger with a large ISP, or acquiring small ISPs with favorable or free peering agreements. Peering will represent a GigE attacker's most significant breakeven obstacle, accounting for slightly more than 30 percent of forward operational expenses.

- **Increase prices**   Some players are currently offering up to 100 Mbps lines for $1,000. Raising prices by 50 percent could reduce the profitability breakeven timeframe.

- **Voice service and *virtual private networks* (VPNs)**   Some players begin offering voice service and IP VPNs as soon as it is technically feasible in order to capture more of a customer's overall telecom bill and benefit from economies of scope.

- **IXCs**   The prospect of entering the metro market gives IXCs an opportunity to strength their backbone operations and develop a new burgeoning revenue stream. Existing customer accounts of the major

IXC players could serve as a significant source of competitive advantage, if they enter the metro transport space as standalone businesses. However, like new entrants, IXCs must execute rapidly and efficiently and cover 10 to 15 metro markets building networks that support efficient scale.

CLECs Top-performing CLECs are positioned to begin offering GigE-based services to augment their existing product lines. They can leverage existing customer relationships through a sales force that's well versed in marketing similar products. CLECs that want to capture this market opportunity could follow one of two paths:

- Invest in the infrastructure on an overlay basis.

- Acquire new GigE entrants inexpensively if they stumble.

Like any other expansion initiative, this choice will depend largely on access to capital as well as protocol/network/operating system compatibility and geographic focus.

ILECs/CAPs Given their assets and customer relationships, ILECs and the large CAPs have the best position to address the metro opportunity. At the same time, they're the most susceptible to market share loss.

From a transport perspective, the ILEC or CAP could preemptively accelerate the build out of SONET-lite and metro DWDM networks to capitalize on existing customer relationships and effectively limit new-entrant penetration. For a typical ILEC, this build out could require as little as $1 billion in capex, a relatively small amount compared to an ILECs' average cumulative capex total of $15 to $18 billion in 2000. Another option, should new entrants gain significant market share, is for the ILECs to consider acquisition. Furthermore, buying an out-of-region new entrant could be a means of jump-starting out-of-region expansion.

From an access perspective, ILECs could attack the market by aggressively offering GigE services over fiber as an add-on service, using (parallel) overlay networks.

Key: If Ethernet as an access service becomes as dominant as predicted, the ILECs may see a major erosion of their DS-3 revenue—possibly to the point where that service/technology may be declining in its product life cycle by 2005 to 2006.

3.13 The Business Approach and Service Delivery Options

A key question across all the available options is which mode of delivery for GigE in the metro area will dominate in the end? Optical fiber? FSO? Lambdas on demand? This question is too simplistic. The research group *New Paradigm Research Group* (NPRG) is convinced that a range of delivery methods will come to coexist for a number of years into the future. This is a sound assessment; the nature of the U.S. economy has proven this techno-business model to be true. The wireless *personal communication services* (PCS) industry is a perfect example. The only given is that fiber will not connect every customer for many years, if ever. It takes time and considerable capital resources to dig up streets or trudge through sewers. It's also just not cost effective to connect everyone with fiber. DSL and soon even higher-speed algorithms being developed to expand the traffic carrying capacity of copper will fill many infrastructure gaps that exist due to geography and economics. FSO appears to be an intriguing development and will certainly connect many customers by 2004 if carriers using that technology compete effectively. However, such platforms are not as simple and inexpensive to deploy as they once were. There's also the issue of what applications and content will be available to justify the cost of fiber deployment with GigE. Many platforms are already available and new ones will be developed once a national IP-based infrastructure is fully in place.

In summary, the MAN GigE-over-fiber rollout is another important step in the larger end run around RBOC networks. Combined with cable, satellite, and wireless, GigE over fiber represents the future of competitive local telecom: an environment in which competitors bridge the last mile with their own facilities or through those of a neutral access provider. Competitive telcos will only be able to defend their territory and achieve sustained profitability through this route. GigE is clearly an important key to that future.

Metro transport will be intensely competitive—it will be highly scale dependent with high concentrations of customers. Survivors other than the ILEC will include just a few new entrants or IXCs that forward integrate. Forward integration means building out metro Ethernet infrastructures in a standalone fashion while simultaneously planning to integrate the new infrastructure into the core of their networks in a phased manner.

3.13.1   Business Model Segmentation

A wide variety of new and incumbent companies are looking to service the explosion in bandwidth demand. Players are pursuing one of four distinct business models:

- Transport service providers that serve data centers and carriers as core transport providers (such as Telseon), and new private metro transport entrants (such as Looking Glass, OnFiber, and Telseon) need to strike deals with anchor tenants to be viable or merge with each other to achieve economies of scale and scope.
- Access service providers looking to serve enterprises (such as Cogent and ILECs).
- Wholesalers of dark fiber
- FSO service providers (such as Terabeam and Tellaire)

These business models differ with respect to their target customers, traffic patterns, technology platforms, sales and marketing approach, and to some degree respective competitive landscapes.

With their broad footprint, existing customer relationships, and access to capital, ILECs are positioned to capture the metro transport revenue opportunity. This may not be a high priority for most ILECs at this time, given larger opportunities in wireless and long-distance services. However, 2002 is seeing more aggressive plans by both RBOCs and IXCs in terms of getting into the metro Ethernet game.

The metro transport market will be a sizzling market because of its unsettled stage of development. This makes it ripe for intense competition. Existing and emerging IXCs (such as Level 3 and Broadwing) could build their own metro transport networks based on their current long-haul backbone traffic volumes. However, such a strategy would involve significant capex for these service providers who already have heavy debt loads. Another option is that they could divert traffic to new entrants to converse cash, potentially securing highly attractive deals through their strong bargaining position.

3.13.2   The Service Providers

Key: While reading through the following list of service providers, it's important to keep in mind that the service provider segment of the telecom industry is in complete turmoil as of 2002. In the highly dynamic telecom industry, things can change very quickly. It's conceivable that any of these service providers could go bankrupt (whether under Chapter 7 or Chapter 11) or could be purchased by another provider. Or in some cases, piece parts of any one service provider could be purchased by another service provider (such as, WorldCom). Many carriers listed here have *already* succumbed to Chapter 11 bankruptcy (such as WorldCom, XO Communications, MFN, and Yipes Communications, to name a few). Many of the *services* listed in this section will morph and evolve over time as well in response to volatility in the industry, economic pressures, competitive pressures, new technologies, new partnerships, and the needs of the marketplace. Some providers will be covered in more detail than others simply due to the nature of their business and service offerings. Some providers will also be covered more than others simply because they're new to the scene or offer service in very nontraditional ways.

The following is an alphabetical listing of the service providers who offer Ethernet products or services. This is followed by another alphabetical listing of the major equipment players in this space (as of 2002).

3.13.2.1 AT&T AT&T unveiled a set of new access and performance enhancements for its Internet services in September 2001, giving businesses more options for using IP networking. In addition to dial-up access from 2,200 AT&T POPs in 60 countries, domestic Ethernet connectivity to the Internet is now available.

AT&T's Metro Ethernet service is point-to-point over SONET, effectively making it an *Ethernet over SONET* (EoS) offering. The service is offered in four port speeds: 50, 150, 300, and 600 Mbps, which map to *synchronous transport signal* (STS) levels. Four implementations are available from the customer premise to AT&T's POP:

- Connections to another customer premise
- Connections to an AT&T local access node
- Connections to another access vendor
- Connections to a long-distance carrier

Although the service is currently offered over SONET transport, AT&T plans to eventually offer the service over DWDM (wavelengths). The service is currently available in the 70 cities where *AT&T Local Network Services* (AT&T LNS) operates as well as in a few additional cities where AT&T uses other CLECs for network capacity.

AT&T LNS has 6,000 office buildings on net in 70 cities with 1.4 million fiber miles deployed. AT&T is also evaluating the potential of serving metro Ethernet customers using fixed wireless technology in some cases, but no Ethernet customers are currently being served using wireless technologies. The rollout of any Ethernet fixed wireless service would be dependent on achieving transmission speeds greater than 100 Mbps, which is the limit of existing hardware.

AT&T Metro Ethernet is offered in addition to *AT&T Transparent LAN service* (AT&T TLS). AT&T TLS operates at 10 and 100 Mbps speeds and is a multipoint architecture, whereas the AT&T Metro Ethernet is point to point.

The AT&T Metro Ethernet service is priced below AT&T's SONET OC-n private-line services and is generally more expensive than the services offered by Ethernet startups. AT&T Metro Ethernet comes with the same SLA offered with AT&T's private-line services. To install the service, AT&T places a multiservice platform at the customer premises, which is also known as an MSPP. This equipment is considered part of the AT&T network and is not the responsibility of the customer.

In addition to AT&T Metro Ethernet services, AT&T also launched an Internet access over Ethernet service in September 2001 called *AT&T Managed Internet—Ethernet Access*. The service will eventually be available in the same locations as the AT&T Metro Ethernet service, 70 AT&T LNS cities plus a few additional cities, and within AT&T hosting centers. AT&T aimed to launch the service in 9 markets in early 2002: Atlanta, Boston, Chicago, Dallas, New York, Philadelphia, Phoenix, San Francisco, and Washington, DC. Internet access is available from 10 Mbps to 1 Gbps in the increments described in Table 3-2.

AT&T Managed Internet—Ethernet Access comes with an SLA that is the same as other AT&T IP-based services, including an installation SLA. After completion of the controlled introduction phase, future enhancements will include *quality of service* (QoS) features and MPLS.

Table 3-2

AT&T Managed Internet—Ethernet Access purchasing options

Interface Speed Range	Increments Available
10–20 Mbps	1 Mbps increments
20–100 Mbps	5 Mbps increments
100 Mbps–1 Gbps	50 Mbps increments

Source: IDC (2001)

The service hands off an Ethernet connection directly to the customer's router. Currently, AT&T IP access over Ethernet customers usually buy one port each, and tend to be former DS-1 and DS-3 Internet access customers.

The Ethernet-based access services are probably a defensive move by AT&T as they see RBOCs and other GigE vendors already active in this space. AT&T's entry into the metro Ethernet market will help them protect some of their high-speed Internet access customer base.

3.13.2.2 BellSouth BellSouth offers a TLAN service called BellSouth *Native Mode LAN Interconnection* (NMLI) service. The tariffed service is offered at 4, 10, 16, or 100 Mbps, although BellSouth currently offers GigE on a special assembly basis. They planned to roll out a tariffed version of the service in the second half of 2001.

BellSouth's NMLI customers tend to be medium-sized and large businesses, and the service has been particularly popular among financial, healthcare, and education vertical markets. A managed version of NMLI is available from BellSouth Communications Services.

In April 2002, BellSouth announced it was strengthening its data portfolio by extending its high-speed metro Ethernet service to businesses, schools, and government agencies in North Carolina and Florida. With speeds up to 1 Gbps, GigE can transmit data 651 times faster than a traditional DS-1 circuit. This type of functionality enables the download of an entire motion picture in 3.5 seconds!

3.13.2.3 Broadwing Broadwing announced plans for a long-haul, point-to-point optical Ethernet service in May 2001. The carrier planned to begin serving its first customers by November 2001. The service would offer Ethernet transport from 25 Mbps to 1 Gbps in 1 Mbps increments. Although IP access and multipoint service will not be offered initially, the carrier intends to offer both features as future enhancements. Broadwing's Ethernet service is being positioned for Fortune 500 customers and is expected to displace DS-1 and DS-3 private lines.

Key: For first-mile access, Broadwing has partnered with Telseon and is also expecting to use leased private lines between the customer's premises and Broadwing POP.

3.13.2.4 Cogent Communications Cogent Communications is based in Washington, D.C., and was formed as a privately held firm by former real-estate executives. Cogent offers retail, Ethernet-based Internet access to customers in MTUs. As such, its initial market entry position was as a BLEC. The carrier uses *Packet over SONET* (PoS) (OC-48) from its POP to the customer premise for its Ethernet transport, running native-mode Ethernet only inside the customer's building. Though Cogent has less of a value-added service story to date, it has earned praise from analysts for its back-office operations support. Cogent is operational in 20 American cities as of 2002.

Although the company's strategy of offering 100 Mbps and 1 Gbps connections to the Internet provides it with quite a competitive advantage over rival ISPs, it is questionable whether the company can sustain itself in the long term by merely providing transport. However, Cogent sources say that their narrowly defined product suite enables them to significantly decrease their sales cycle. Cogent is part of the ELEC story and will be a likely prospect for offering 10 GigE once it's standardized.

Key: Cogent's Ethernet-based Internet access service is offered in increments of $1,000 per 100 Mbps and up to $10,000 for 1 Gbps of Internet access. This is extremely disruptive pricing, and the entire telecom industry has taken notice of this low-priced offering. Some industry analysts call this irrational pricing that is used by Cogent in an attempt to obtain market share quickly. Time will tell if this approach truly hails a paradigm shift in transport pricing models for the telecommunications industry or if it's simply too aggressive for an upstart carrier to sustain over time and still be profitable.

The installation fee for Cogent's service described previously is $10,000. No *customer premises equipment* (CPE) is required of the customers, and the service is delivered via a standard Ethernet jack (RJ-45).

3.13.2.4.1 *Target Segment* Cogent's typical customer has approximately 30 desktops. Many are law firms or financial services companies where Cogent service tends to replace DS-1-based Internet access. Cogent has quite a competitive advantage in that it is able to offer customers roughly 65 times the bandwidth they currently get from their T1 connection at a price equal to or often times less than what they are paying now. How can Cogent do this? Well, that's where the true competitive advantage comes into play. By purchasing fiber from Williams Communications (for long haul), MFN, and Level 3 (metro area), and using Cisco equipment, Cogent has created an end-to-end optical network that enables the company to utilize Ethernet over DWDM. This type of architecture provides an enormous amount of bandwidth at a relatively low cost.

The demand for Cogent's service has outpaced the original parameters of the network, but because the company uses DWDM, it can easily add more capacity by activating additional wavelengths. Cogent's network now has 80 Gbps of capacity and continues to scale as needed. The architecture that Cogent chose for its network also enables the company to provide SLAs that are on par with industry-leading ISPs.

Key: Cogent's next claim to fame is that it will provide all of this on a nonoversubscribed network. In an industry where an over subscription rate of 400 percent is common, this is an exceptional strategy.

Cogent has national network ambitions and leases fiber-optic capacity from companies such as Williams Communications. Cogent has won considerable mindshare with its unbelievable sales pitch to provide Internet connections at speeds 100 times faster than conventional T1 data lines for the same price.

In July 2001, Current Analysis reviewed Cogent's market status and assessed that the company's initial strategy of a slow expansion to build a strong brand name and customer base appeared to have paid off. Cogent has seen so much increased demand that it had to upgrade its network and planned to expand into 14 cities by the end of 2001.

On the flip side, it's important to remember that this company basically serves a niche market: business customers located in multitenant buildings in large urban markets. To narrow the niche even more, Cogent requires that the buildings it serves are located within 1,250 feet of the hubs of their network. Businesses and buildings not fitting these criteria are pretty much out of luck. Even if Cogent does deem that a building and its tenants meet its criteria for providing service, it can take up to nine months to turn up the first customer in a new building.

Cogent's purchase of metro fiber in 18 cities from Level 3 and access to Level 3's data centers in July 2001 is a positive sign because Cogent has so far taken a cautious approach to expansion; instead, it has concentrated on building a customer base and a strong brand name. More than anything else, Cogent's expansion is a sign that the company has had great success and is now in a position to provide its service to all 18 cities touching its backbone network. These markets include Atlanta, Boston, Chicago, Dallas, Denver, Houston, Los Angeles, Miami, New York,

Newark and Weehawken, New Jersey, Philadelphia, San Diego, San Jose, San Francisco, Seattle, Tampa, and Washington, DC. This agreement also provides a nice complement to the metro area fiber the company purchased earlier from MFN. Although Cogent's purchase will not have an immediate impact on the Ethernet services market, it will have a moderate to high impact by late 2002, as Cogent begins presale efforts to light these cities. It was important for Cogent to negotiate access into Level 3's data centers in this agreement as ELECs like Telseon are very active in this area.

Signing an agreement with Level 3 for metro area fiber was the easy part. Cogent must still build fiber laterals to the buildings it wants to serve, which means getting permits, digging up streets, and actually building to those locations.

Cogent follows an extremely strict set of requirements to ensure that its network is not oversubscribed. No more than eight buildings exist on a metro (access) ring, and they will not sell more than the capacity of the ring. As demand increases, the company drops in another wavelength rather than overprovisioning existing connections. Rather than buying more dark fiber as demand begins to exceed capacity, the company uses DWDM to scale its network, which greatly reduces the time it takes for the company to provision more bandwidth to a specific building.

3.13.2.4.2 Cogent Purchases Defunct BLEC Allied Riser Communications (ARC)

On February 4, 2002, *Allied Riser Communications* (ARC) was acquired by Cogent Communications. ARC offers broadband access, VPN, and other services over Ethernet to companies located in MTUs. Seventy percent of ARC's customers were small businesses with an average of 25 employees. Cogent formally merged with ARC at the end of August 2001, but it appears that Cogent only bought ARC's in-building networks—not its access rights. The companies would not reveal details of the merger. ARC brings 900 buildings in 40 markets to Cogent's fingertips. In addition, ARC was known for going into older buildings such as the Empire State Building and replacing copper infrastructure with fiber. But it's safe to say that even with ARC's in-building networks, Cogent will still have to lease or build out metro rings and last-mile fiber.

One of the key benefits to Cogent's acquisition of ARC is that ARC's in-building networks should drive traffic onto Cogent's backbone, which will increase network efficiency and profitability. The difficulty for Cogent is that customers are reeling from ARC's failure, and Cogent will have to convince them of its own stability.

Key: Cogent still needs a transport bridge to go from the metro core to the access core, and that involves laying fiber. This is expensive and time consuming.

3.13.2.4.3 Cogent Purchases Failed ISP PSINet Also in February 2002, Cogent announced its purchase of PSINet's U.S. network and customer assets. Cogent plans to purchase PSINet's major U.S. operating assets, which include a portion of the company's customers, network, equipment, and three hosting centers.

PSINet filed for bankruptcy protection under Chapter 11 in June 2001 after being delisted from Nasdaq. The stock market delisted the company's stock in April 2001 shortly after PSINet announced job cuts and a fourth-quarter 2000 loss of $3.2 billion. Although PSINet still has some international assets, the majority of the once high-flying ISP will be dissolved if the bankruptcy court approves Cogent's acquisition.

Cogent's other advantage over metro fiber competitors is that it owns a nationwide backbone, not just a local loop. Cogent's network build out was made possible in part by a $280 million Cisco partnership and dark-fiber leases of $100 million from MFN and $215 million from Williams Communications. See Figure 3-5 for a map of Cogent's national network footprint.

Figure 3-5
Cogent's national network

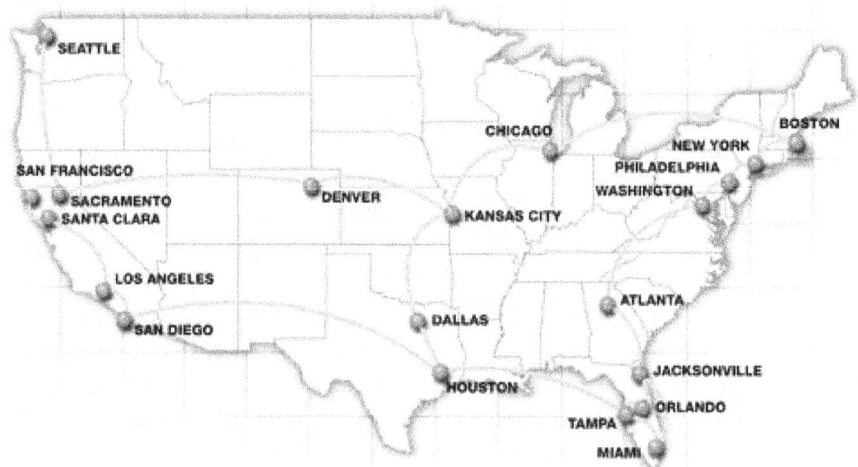

3.13.2.5 Cypress Communications Cypress Communications delivers Internet access over Ethernet to customers located in MTUs in seven markets: Atlanta, Dallas, Boston, Houston, Chicago, Seattle, and Los Angeles/Orange County. The company has a narrow market focus, targeting upscale office buildings that charge more than $34 per square foot. The average Cypress customer has 15 to 20 desktops and is a professional services firm, such as an accounting or law firm.

In Seattle, Cypress owns its own fiber in the metro area and uses Ethernet transport between the customer locations and the Cypress POP. In other cities, Cypress uses primarily copper-based T1 transport.

Cypress offers a standard SLA for availability. Other SLAs are offered on an *individual case basis* (ICB). Pricing for Cypress' service is primarily flat rate, with some measured service in Seattle to reflect the company's ownership of its own fiber in that area.

3.13.2.5.1 Commercial Building Access Rights Cypress Communications has executed long-term license agreements with some of the nation's most well-renowned commercial property portfolio owners. These licenses give Cypress Communications the right to install, operate, and maintain its fiber-optic networks in more than 1,300 buildings throughout the United States. The agreements also provide for certain in-building sales and marketing privileges. Cypress Communications, Inc. is a now wholly owned subsidiary of U.S. RealTel, Inc. (In February 2002, U.S. RealTel completed the acquisition of Cypress Communications, Inc.)

3.13.2.6 Everest Broadband Everest Broadband Networks is a BLEC in the *multitenant broadband service provider* (MBSP) marketplace. Everest caters to small to mid-sized businesses that occupy multitenant buildings and hotel properties. The company focuses on the business traveler. Everest offers telecommunications services such as high-speed Internet access, local and long-distance telephone service, and web applications. Everest delivers these services over a metropolitan Ethernet service architecture.

Although Everest focuses on retail service, the carrier also sells Ethernet transport on a wholesale basis to other carriers. Everest has purchased its IP backbone services from the Global Crossing (which filed Chapter 11 in 2002) and Level 3, and it has signed contracts for fiber with MFS Communications and Telseon as well. A typical Everest customer has 25 workstations and is migrating from DSL, T1, or dial-up Internet service.

Everest places a layer 2 or 3 switch and router in the basement of each of its on-net buildings and runs fiber up the building risers to serve its customers. It also operates optical Ethernet rings in each of its seven markets: New York, Los Angeles, Dallas, Houston, Montreal, Toronto, and Washington, D.C. Most of its on-net buildings are currently connected to its metropolitan network via copper T1s as Everest waits on fiber build outs to be completed. Everest offers SLAs on bandwidth, network availability, and installation in lit buildings.

3.13.2.7   Giant Loop   Giant Loop is a BLEC-based in Waltham, Massachusetts. It was founded by ex-EMC executives (EMC is a major U.S. manufacturer of storage devices and components). Giant Loop focuses on the Global 2000 businesses. The company primarily targets the largest financial institutions in money-center cities globally. Giant Loop sells outsourced telco services as well as storage network services.

Giant Loop has field offices at the following locations: New York, Philadelphia, Pittsburgh, Chicago, San Francisco, Seattle, and London. Unfortunately, the company's web site (May 2002) leaves something to be desired in terms of concrete company information, serving areas, and service offering details. Their web site does not have a *frequently asked questions* (FAQ) link, which is sorely needed.

3.13.2.8   IntelliSpace   In 1995, IntelliSpace created the world's first smart building, which played a part in the broadband revolution. In August 1997, IntelliSpace completed construction of New York's largest privately owned MAN. In March 2000, IntelliSpace completed construction of the world's largest MAN. In May 2000, the company lit their 300th building: Carnegie Tower. As of October 2001, IntelliSpace is serving over 60,000 business users with over 1,000 lit buildings in over 150 cities and communities around the world. Some of IntelliSpace's more notable business successes have been developing intelligent, broadband-ready buildings in the following locations: Rockefeller Center in New York, the Jacob Javits Convention Center and the Lincoln Center in New York, the Chicago Board of Operations Exchange in Chicago, Old City Hall in Boston, the Bourse Building in Philadelphia, and Tower 42 in London. Figure 3-6 illustrates a typical IntelliSpace network architecture.

IntelliSpace offers e-mail via broadband access, videoconference, and other services over Ethernet to companies located in MTU office buildings in 13 U.S. markets. Sixty percent of IntelliSpace's customers are small companies or branch offices of large companies. The remaining 40 percent of its customers are large-enterprise customers. IntelliSpace offers port speeds from 64 Kbps to 10 Gbps, and 60 percent of its ports are less than 1 Mbps.

IntelliSpace connects customers to its on-net buildings using a combination of fiber, copper, and wireless, although most buildings' connections

Figure 3-6
IntelliSpace network
model

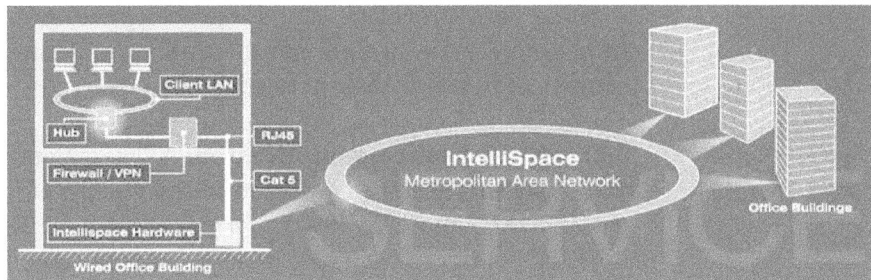

Figure 3-6
IntelliSpace network
model

are via T1. Within each building, IntelliSpace deploys optical Ethernet in the riser.

IntelliSpace operates a 10 Gbps ring in New York and New Jersey using Extreme Networks equipment and has signed a $215 million fiber deal with MFS. The company is also acquiring fiber from Telergy and other utilities.

3.13.2.9 Looking Glass Networks Looking Glass Networks is a facilities-based carrier headquartered in Oak Brook, Illinois (a Chicago suburb). The company builds, owns, and operates metropolitan optical networks offering a broad range of data transport services to both carrier and enterprise customers in Tier 1 American cities.

Looking Glass was formed by telecommunications executives with a proven track record in the deployment and management of successful metro networks (they're former MFS executives). Since its launch in April 2000, Looking Glass has raised $475 million in total capital including $200 million in initial private equity financing.

Looking Glass has franchise and right-of-way authorization to build their networks to connect major aggregation points (for example, carrier hotels, ILEC COs, and MTUs) in the largest U.S. metro areas. They are deploying physically diverse, high-capacity optical networks in the following cities: Atlanta, Boston, Chicago, Dallas, Houston, Los Angeles, New York/northern New Jersey, San Francisco/San Jose, Seattle, and the Washington D.C./northern Virginia area.

Looking Glass' end-to-end, redundant networks offer the full spectrum of SONET, Ethernet, and DWDM (wavelength) services on application-neutral optical platforms. Their primary product offerings include SONET (FastGLASS[SM]), wavelength (WaveGLASS[SM]), and Ethernet-based lit services (EtherGLASS[SM]) that support traditional and nontraditional applications at speeds ranging from 10 Mbps to 10 Gbps. Additional offerings include high-capacity dark fiber and carrier-neutral collocation services.

Looking Glass' target customers include IXCs, ILECs, CLECs, ISPs, data centers, bandwidth-trading organizations, storage facility providers, wireless data providers, and large-enterprise customers. The company turned up service in nine metro markets in January 2002, using Cisco equipment to power their network.

3.11.2.10 Qwest Communications Qwest Communications is a company that was formed as a result of the merger of Qwest (a former independent IXC) and U.S. West (a former Baby Bell). Qwest no longer offers its TLAN service to new customers, and existing month-to-month customers could only continue to subscribe to the service until August 30, 2001, when the service was formally grandfathered. Qwest's TLAN service was offered in speeds of 1.544, 4, 10, or 16 Mbps and interconnected either Ethernet or Token Ring LANs. The maximum TLAN circuit distance was 75 miles. A multipoint feature enabled customers to connect up to 15 TLAN data links of the same data type and speed. An inter-ring link feature connected data links between service points. The maximum length of an inter-ring link is 25 miles. TLAN services have been available in the contiguous United States and on an intraLATA basis only in Arizona, Colorado, Iowa, Idaho, Minnesota, Montana, Nebraska, New Mexico, North Dakota, Oregon, South Dakota, Utah, Washington, and Wyoming.

In May 2001, Qwest announced a new Ethernet service to compensate for the grandfathering of its TLAN service, which was offered by the unregulated side of the company called *Dedicated Internet Access over Ethernet*. The service offers speeds from 1 Mbps to 1 Gbps and was initially launched in Chicago, Dallas, Los Angeles, New Jersey, New York, San Jose, and Washington, D.C. Qwest is initially offering the service to customers who subscribe to its data center service, although the carrier also plans to target enterprise customers. Service is provided over Qwest's own fiber in the 27 cities where it is building metropolitan fiber networks as well as over Telseon fiber through a partnership agreement between the two carriers. Qwest offers an SLA for network availability, and customers can change bandwidth using a web-based interface.

3.13.2.11 SBC Communications SBC's initial foray into the metro Ethernet market occurred in July 1999, when its Ameritech business unit launched a point-to-point GigE service called GigaMAN^SM. GigaMAN is offered on a tariffed basis by the regulated telco.

SBC expanded the popular GigaMAN offering to the other Bell regions in January 2001. The service was previously limited to a distance of approximately 31 miles, but in mid-2002, SBC launched a mid-span repeater enhancement for GigaMAN service. This enhancement allowed for the extension of GigaMAN service to a distance of approximately 60 to 70 miles end to end, depending on the number of splice points in the circuit path and the age of the optical fiber.

In October 2002, SBC signed a contract with Nortel Networks to purchase Nortel's Optera 5100 system for use in its network. SBC plans to use the Operta 5100 to expand its managed wavelength service offerings, including optical Ethernet and storage services, throughout its 13-state territory.

SBC is also considering and/or developing a switched Ethernet service, a fibre channel service, and a 10 GigE service. An Ethernet-over-sonet product is also under development.

3.13.2.12 Telseon   ELEC Telseon, based in Denver, offers Ethernet services in 20 U.S. markets: Atlanta, Chicago, Cincinnati, Dallas, Denver, Detroit, Houston, Los Angeles, Miami, New York, northern Virginia, Orlando, Philadelphia, San Diego, San Francisco, Seattle, San Jose/Silicon Valley, St. Louis, and Tampa.

Telseon provides optical Ethernet transport between data and collocation centers within metropolitan areas over metropolitan DWDM. They mainly offer Ethernet connectivity in the metro core—they're more focused on Ethernet *transport* versus Ethernet *access*. In this way, they're operating in almost the opposite mode of Cogent. Telseon is a more wholesale-oriented provider that interconnects data centers, collocation, MTU facilities, and carrier hotels within a metro largely serving other providers already located at one or more of Telseon's POPs. They now have 100 data and/or collocation centers in 20 NFL cities (Tier 1 metros) and at least 300 customers.

3.13.2.12.1 *The Telseon Backbone and Services Portfolio*   Telseon wholesales backbone connectivity from five different carriers (Level 3, 360 Networks, Broadwing, Verio, and Dynegy), and customers can choose one based on quality assurances or a particular core competency. They can also access higher-level, value-added services from these backbone providers by ordering them through Telseon, which acts as the retail agent and delivery mechanism for a broad portfolio of applications and services. Telseon launched its Ecosystem Web Directory of these services in April 2001, which includes more than 50 possibilities, such as caching, collocation, data management, disaster recovery, enhanced storage services, *enterprise resource planning* (ERP) applications, file sharing, managed database services, managed modems, network design, security services, site monitoring, streaming media, videoconferencing, VPNs, and web hosting. Telseon's target market segments are

- ASPs
- ISPs
- Large enterprises
- CLECs
- IXCs

The carrier has purchased fiber from Level 3, MFN, and XO as well as municipalities and other suppliers to build their network infrastructure. Telseon's customers are ISPs, storage providers, carriers, and content providers. They include Broadwing, Everest Broadband, and Qwest among others. The company offers bandwidth from 1 Mbps to 1 Gbps, with 50 Mbps being the average. From 1 to 20 Mbps, bandwidth is available in increments of 1 Mbps. From 20 to 100 Mbps, the increment is 10 Mbps, and between 100 Mbps and 1 Gbps the increment is 50 Mbps. Telseon customers can use a web-based provisioning interface to change bandwidth on demand. Telseon offers SLAs to its customers, which cover network availability, packet loss, *mean time to repair* (MTTR), and latency based on two *classes of service* (CoSs). Network availability is guaranteed at 99.99 percent, with an option to upgrade to 99.999 percent. Installation and bandwidth provisioning SLAs are also available. Telseon uses GigE principally in cities where it acts as a speedy distributor, connecting various data centers. Although the upstart in Englewood, Colorado, avoids the headaches of last-mile access and the haggling over equipment space with the Bells, its fortunes are tied to soon-to-be-dismantled customers such as NBC Internet. Telseon's brass is a collection of CLEC veterans, notably including two execs from MFS Communications (Mike McHale, the Chief Operating Officer, and Sean Whelan, the Vice President for Strategic Alliances). (MFS was one of the first competitive access providers in the game back in the 1980s and 1990s.)

3.13.2.12.2 *Sticking to the Game Plan* Telseon is one provider that has so far stood by its mantra of partnering for last-mile access rather than buying a BLEC's assets like Cogent has done. On July 31, 2001, Telseon partnered with Everest Broadband Networks so that Everest customers in four markets would have access to Telseon's IP-based network. Everest will be able to wholesale bandwidth to other Telseon service provider customers who might want to reach MTU enterprise customers.

Telseon remains focused in the metro core rather than the metro edge, because of the "enormous amount of risk at the edge," says Michael Hulfactor, Telseon's Director of Market Analysis. "Many of the tenants in the tall shiny buildings are low-bandwidth users. Lots of businesses are looking for ISP connections, but they are not going to be buying value-added services."

In addition to having preexisting contracts with local incumbent carriers, many businesses in MTUs are branch offices, where major decisions are not made. Telseon is also looking at alternative last-mile technologies, but rather than investing directly in them, the company would partner with other service providers. Telseon has partnered with FSO vendor Terabeam and is considering partnerships with other players in the wireless and GigE-over-copper space as well.

Telseon has an aggressive, optimistic deployment schedule. Telseon has become a dominant metro wholesale and retail Ethernet provider, and their plan was to deploy 10 GigE by the end of 2001, with a slow upgrade in 2002. This puts a lot of faith in the first generation of 10 GigE gear!

Telseon is also confident in the demand for 10 GigE as an aggregation tool for Gbps links and WDM traffic. Both can be converted to a clear-channel 10 GigE link. The company wants guaranteed interoperability, but won't require any new functionality or provisioning capabilities from vendors and doesn't anticipate any particular CPE issues or concerns, according to Vipin Jain, Telseon's Vice President of Systems Engineering. QoS is expected to equal, not surpass, GigE benchmarks as currently defined—a murky area for Ethernet providers right now, especially when running more delay-sensitive applications.

In April 2001, Telseon announced its launch of Metro Wave Service, which offers high-capacity, protocol-independent, DWDM-based connectivity solutions to carriers and service providers. The service gives customers connectivity between two network locations within a given metropolitan market, transporting data and IP traffic at 2.5 or 10 Gbps speeds.

The industry's perspective on Telseon's Metro Wave Service was lukewarm when the service launched in April 2001, because at that time there was little demand for 2.5 or 10 Gbps metro wavelength services. Telseon was a year or two ahead of the demand curve. This is an embryonic market that won't see significant growth until 2003, as the cost of bandwidth continues to slide. Due to the high capex associated with deploying new services, most network service providers usually cannot get away with deploying a product ahead of the demand curve. Telseon seems to have circumvented this trap by entering into a five-year network service agreement with Dynegy Global Communications as a way to offset a portion of the capex associated with developing and deploying this product. The market impact of this service launch is low because at this time very few applications require this much bandwidth. Only a very limited customer segment has a legitimate need for 2.5 or 10 Gbps point-to-point connections.

Telseon's Metro Wave Service is designed primarily for carriers and large ISPs, providing them with 2.5 or 10 Gbps point-to-point connections between two or more of Telseon's collocation POPs within a given metro area. The service is protocol agnostic, supporting most high-speed data protocols, and it can be turned up in a matter of days or weeks. Telseon's plans called for rolling out the service in June 2001. Initial markets appeared in Atlanta, Chicago, Dallas, Denver, Los Angeles, Miami, New York, San Francisco, Seattle, and Washington, D.C.

Telseon's Metro Wave Service is a logical addition to the company's service portfolio. Coupled with Telseon's GigE services, which provide customers with metro area connections ranging from 1 Mbps to 1 Gbps, the higher-speed wavelength service now gives the company the ability to target the full gamut of carriers and ISPs. This widens Telseon's target segment to include high-end customers and also provides a new revenue stream for the company. The key to Telseon's value proposition is that all of the company's product offerings provide the customer with high-speed connections at a much lower cost. GigE services typically provide customers with roughly twice the bandwidth at half the cost. Telseon claims that customers of its Metro Wave Service will save 35 percent over comparable OC-n wavelength services.

Competitors employing SONET-based metro networks will be quick to point out that SONET networks, by providing automatic rerouting in the event of a network outage, have built-in redundancy. Wavelength service, on the other hand, has no built-in redundancy feature; therefore, the customer is required to purchase a second DWDM circuit on a separate physical path to provide backup.

Prior to the announcement of its Metro Wave Service, Telseon's service portfolio was specifically aimed at the small to mid-sized carrier/service provider. The company needed to add this high-speed wavelength service to fill out is product portfolio and target the more lucrative large carrier customers.

Up until now, Telseon's network only supported IP traffic. In reality, data is transported using a variety of protocols such as TCP/IP, frame relay, Ethernet, and so on, with no one data protocol clearly winning out over another. The company needed to add a service that is protocol agnostic. Telseon can provision an OC-48 or OC-192 wavelength across its MAN and turn it up in a matter of days or weeks after receiving an order—a significant time saver compared to other network construction options.

Telseon's network services agreement with Dynegy expands Dynegy's network footprint into Telseon's 20 markets and data centers therein while providing Telseon with a partner to shoulder the significant capex costs associated with deployment of the product.

Although Telseon is expanding rapidly—from 60 data centers in December 2000, to over 100 data centers by May 2001—the company's fiber network is still limited. Customers in need of these ultra high-speed connections to data centers that are not on Telseon's fiber path will have to wait for the company to complete its expansion. The need to construct fiber laterals where necessary may add to this delay.

Startups such as Giant Loop and Sphera will need to follow Telseon's lead and quickly expand into key cities. The longer it takes these companies to move into cities such as Atlanta, Chicago, Los Angeles, Dallas, and so on, the more market share they lose.

With Telseon service, on-net customers can add or change logical connections as well as their bandwidth (from 1 Mbps to 1,000 Mbps to 1 GigE) with the Telseon IP Provisioning System, the most advanced in the industry to date. Its latest version, 2.0, is unique because it enables customers to set up new connections directly to any other business on Telseon's network, a huge boost to *business-to-business* (B2B) applications.

Telseon has made security a major focus of its architecture, not just an add-on. Features include leakproof connections—connections that guarantee that data is transmitted only between authorized sources and destinations, intrusion protection, spoofing lockout, network intrusion protection, physical security, admission control, delivery protection, and *Secure Sockets Layer* (SSL) encryption and authentication. Telseon SLAs are also strong and cover packet loss, bandwidth-on-demand guarantees, on-time installation, MTTR, and packet delays.

3.13.2.13 Tellaire Tellaire, a privately held Dallas corporation, is an emerging local loop service provider offering rapid network deployment of advanced, laser-based broadband communications (FSO). Tellaire employs highly secure and reliable fixed wireless optical laser technology to address customer needs to quickly implement high-bandwidth data services in the last mile at a favorable cost structure.

Tellaire aims to provide wholesale last-mile service at speeds from DS-1 to OC-12, and eventually at 1 Gbps. It relies on wireless, nonmicrowave technology for service provisioning. The company lays a mesh of *radio frequency* (RF) and laser links over fiber optics, connected to *Internet data centers* (IDCs), carrier hotels, and MTUs. The key benefit with this strategy is the complete bypass of the ILEC CO. Tellaire is profiled more extensively later in this chapter.

The company announced the availability of a first-of-its-kind, laser-based communications network in October 2000. The service is designed to bring broadband services to the last mile for commercial building owners, BLECs, and ISPs.

Tellaire's unique approach makes it the only wholesale local loop service provider using the most advanced wireless technology to solve the last-mile challenge. Offering speeds up to six times faster than today's best broadband wireless systems (and up to 100 times faster than DSL), Tellaire's solution is configured for 99.99 percent availability. In addition, because the technology has military origins, the laser transmissions are highly secure.

Tellaire's network solution involves placing small laser transceivers on building rooftops and connecting them to routing equipment installed in the buildings. Wireless network traffic is routed from one laser to another via the company's redundant mesh network and ultimately to the Internet. Since it is fully deployable within days of gaining rooftop access, the Tellaire network gives commercial property owners, BLECs, and ISPs the ability to quickly implement high-bandwidth, highly reliable data services in the last mile for their tenants and customers. Tellaire's solution is currently being used for high-speed communications between multiple locations in the Washington, D.C. area by a leading national ISP.

Key: During the months this ISP has used the laser-based solution, the technology has proven extremely reliable—maintaining continuous availability even through heavy thunderstorms.

Additionally, Tellaire has reached agreements with Crescent Real Estate Equities Company to provide service for certain buildings in its Austin and Houston portfolios and with Vornado Realty Trust in New York to begin offering services to the wholesale providers that serve their buildings.

Tellaire's technology partner is Perot Systems Corporation, a worldwide provider of information technology services and business solutions. Perot Systems provides the back-office technology infrastructure to facilitate delivery of Tellaire's network services to its customers. Tellaire also has an existing partnership with *Wireless Facilities Inc.* (WFI) to provide 24×7 network monitoring and management services through its state-of-the-art *network operations center* (NOC) in Dallas.

3.13.2.14 Terabeam Terabeam is based in Kirkland, Washington, and offers fixed wireless Ethernet-based services using a 16-inch receiver that is installed in an office window. Terabeam Internet Connection offers 5, 10, or 100 Mbps MPLS-enabled IP-over-Ethernet connectivity, and Terabeam Metro Connection uses 10 and 100 Mbps IP over Ethernet for LAN interconnection.

Terabeam provides high-speed, last-mile broadband IP services using its proprietary FSO laser technology. It is currently creating a network using its Fiberless Optical™ solution over a MAN to connect LANs and WANs.

Terabeam launched its service in Seattle in March 2001 with four customers. As the world's premiere provider of FSO technology, Terabeam designs, manufactures, sells, and supports carrier-class systems that enable service providers to extend the reach of their fiber networks, improve customer service, and enhance ROI. Terabeam FSO systems are known for their high performance, reliability, ease of installation, and lower *total cost of ownership* (TCO). Terabeam also offers unique and innovative installation tools and services, and prides itself on outstanding customer services.

Some of the benefits of making Terabeam's FSO technology part of a provisioning toolkit are as follows:

- It allows for reaching locations where fiber is not technically or economically viable.
- It is less expensive than installing and lighting private fiber.
  - No expensive and time-consuming trenching is required, along with the need to obtain access to all the necessary rights of way and permits, and the need to manage the project.
  - It's a cost-effective way to complete metro networks.
- FSO is a less expensive option than other wireless technologies.
  - No *Federal Communications Commission* (FCC) or other governmental permits are required, as the frequencies used are unlicensed spectrum. There's no requirement to buy spectrum (similar to the PCS license auctions in 1995).
  - FSO is a simple, through-the-window transmission system.
- These systems purport to have infinite scalability.
- They have the ability to deploy rapidly.
- They can be easily integrated into existing networks.
- They have lower TCO than other FSO solutions.
- They have unique installation tools.

Terabeam also offers consulting and training services to assist carriers and enterprises with the deployment and use of FSO.

Terabeam has assembled the largest suite of FSO development labs anywhere in the world. Their scientists and engineers have been awarded three patents and have more than 60 patents pending. Future generations of Terabeam equipment will deliver even greater performance at reduced costs.

The following are key features of one of Terabeam's primary products—the Elliptica 7000 transceiver.

- The Elliptica is the only *American National Standards Institute* (ANSI) and *Interexchange Carrier* (IEX) Class 1 high-performance FSO system in the world today.

- It incorporates automatic, patent-pending, high-speed pointing, and tracking.

- It supports both Fast Ethernet (100 Mbps) and OC-3/STM-1 (155 Mbps) protocols.

- It installs quickly and automatically aligns.

- The Elliptica maintains high availability with continuous self-optimization to counter building sway and other vibrations.

- It uses a built-in video camera and remote access for fast, convenient alignment (when necessary) and remote link monitoring and management.

- It has an ANSI and IEC Class 1 eye-safe transceiver for maximum safety under any viewing conditions.

- The small beam provides more receive power, higher availability, enhanced security, and longer links.

- The integrated network management system provides continuous configuration, performance, security, safety, and fault management.

- This FSO system can be installed indoors or outdoors.

Figure 3-7 shows a front view of Terabeam's Elliptica 7000 series tranceiver. Note the small footprint.

3.13.2.15   Time Warner Telecom (TWT)   TWT began as a partnership between Time Warner and US West, and experienced rapid growth through 2001. Since early 2001, TWT has increased its presence from 24 to 39 markets, thanks in large part to its purchase of most of the assets of bankrupt provider GST Telecommunications. As of 2002, TWT is active in 44 cities. Even though Time Warner still owns 47.8 percent of TWT, the firm is independently run. Last year, Time Warner generated more than $487 million in revenue.

Figure 3-7
Terabeam's Elliptica
7000 FSO transceiver

TWT offers 10 Mbps, 100 Mbps, and 1 Gbps Ethernet service within metropolitan areas. TWT targets Fortune 2000 businesses, generally those with more than 300 employees. The service is primarily used for TLAN and currently uses a point-to-point architecture, but the carrier is exploring use of a multipoint technology for possible launch in the third quarter of 2001. LuxN and Cisco provide equipment for the TLAN service.

TWT delivers GigE and other high-speed services via its new DWDM platform. The carrier recently agreed to spend an unspecified sum on LuxN's WavSystem DWDM equipment, which lets providers push up to 16 channels over a fiber pair. The LuxN gear will enable TWT to provide any service directly over DWDM without requiring SONET or additional fiber. The end result will be quicker provisioning timeframes. The DWDM equipment will also let providers increase capacity on nets in areas where bandwidth is running short.

TWT plans to use the LuxN gear to roll out OC-3 (155 Mbps) to OC-48 (2.5 Gbps) services in addition to GigE. These offerings would compete with high-end data services from traditional carriers (ILECs and IXCs) as well as newcomers including Cogent and Telseon.

Because the underlying DWDM technology is protocol independent, TWT can also offer services such as Fibre Channel over DWDM for traffic routing to and from data centers. Ultimately, TWT plans to use DWDM to offer bandwidth on demand (lambdas on demand). TWT already offers a suite of traditional services including DS-1, DS-3, native LAN, SONET, and voice.

TWT has managed to become one of the few profitable CLECs in the United States, according to Jeff Moore, an analyst with research firm Current Analysis. "One of the keys to their success is that they're facilities-based," he says. "They have a lot of fiber, 21 Class 5 switches, and 21 markets with data centers." Although TWT operates in 15 Tier 1 cities, Moore notes that the provider also serves a large number of Tier 2 markets, such as Greensboro, North Carolina; Memphis, Tennessee; and Columbus, Ohio—where the competition isn't as fierce. See Figure 3-8 for an illustration of TWT's nationwide network.

3.13.2.16 Verizon Communications Verizon Communications is a company that's the result of the merging of three huge incumbent telcos: Bell Atlantic, Nynex (another former Bell company), and the subsequent merger of those two Bell companies with the former largest independent telco in the United States—GTE.

Verizon offers an Ethernet TLAN service at speeds of 10 and 100 Mbps on a tariffed basis in Maryland, Virginia, New York, and Washington, D.C.,

Figure 3-8
TWT's national
network

TIME WARNER ⬡ TELECOM

and on a special-assembly basis throughout the rest of Verizon's territory. Verizon also offers 1 Gbps TLAN service on an ICB only.

Key: Basic TLAN service is available to customers located up to three miles from a Verizon CO that's equipped with TLAN equipment. An additional charge is applied to provide service to customers located between 3 and 15 miles away from a TLAN-equipped CO, but the service is not available to customers located more than 15 miles from the CO. For 100 Mbps access, all customers' locations must be served by the same CO.

In July 2001, Verizon expanded beyond its local serving area in the Dallas/Fort Worth Metroplex by building six fiber-optic rings in conjunction with MFN that cover 270 route miles. With this expansion, Verizon's *Enterprise Solutions Group* (ESG) is targeting large business customers with pure data transport using DWDM and SONET. Switch-based services such as local and long-distance voice, frame relay, and ATM will be offered at a later date.

Verizon entered the Dallas area as an optical-based CLEC in the summer of 2001, competing with incumbent SBC and other rivals. The company planned to enter into two additional out-of-territory markets by the end of 2001—Seattle and Los Angeles.

Unlike many competitors in the telecom field that have tried to expand into the heart of another RBOC region, Verizon is well positioned to effectively compete against out-of-region incumbents. The company will only enter markets where it can quickly provide customers with a complete communications solution and where it has already established a strong brand name. How is Verizon able to do this? They have relationships with both MFN and Genuity for dark fiber and transport. More importantly, Verizon has the assets it acquired during the Bell Atlantic/GTE merger. GTE, formerly a huge, independent (non-Bell) ILEC in many areas throughout the country, typically served suburban areas of larger cities. Verizon is taking the present GTE established around the Dallas/Fort Worth area and pushing it deeper into the business district in order to provide on-net services to a more lucrative customer base—large businesses. With its financial backing, strong brand name, and suite of communication solutions, Verizon provides business customers with a legitimate alternative to SBC. This may be reassuring at a time when CLECs are dying, and the IXCs are divesting and spinning off operating units in an attempt to reposition themselves as high-growth companies.

Verizon has the resources and reputation to quickly gain market share in the lucrative large business market. If and when it is permitted to reacquire Genuity (which it spun off as a condition of the Bell Atlantic–GTE merger), Verizon will be a powerful national player with control of its own backbone throughout the United States. If SBC is unsuccessful in its efforts to offer interLATA services by that time, it could lose a significant amount of national and multinational customers to Verizon.

3.13.2.17 WorldCom WorldCom was the nation's number-two long-distance carrier, but the company will be known forevermore for its $3.8 billion accounting fraud that was uncovered in July 2002. Several weeks after the fraud was uncovered, WorldCom declared Chapter 11 bankruptcy. Nevertheless, a review of the company's current metro Ethernet offerings is called for because somehow, in some way, the company and/or its piece parts will survive the crisis. WorldCom will emerge from bankruptcy in part or in whole. It's very possible an RBOC will buy one or more of the troubled carrier's business units. To that end, services listed here may also survive in some form, in some way. This includes WorldCom's metro area Ethernet offerings.

In May 2002, WorldCom announced a nationwide metro Ethernet service. They rolled out a series of services based on point-to-point Ethernet connections and claimed this was just a start. The services include site-to-site links as well as corporate links to the Internet and enterprise VPNs.

WorldCom is now the third largest incumbent carrier offering Ethernet services. BellSouth offers a series, as does SBC Communications. Table 3-3 shows a listing of WorldCom's metro service offerings.

WorldCom is basing its Ethernet offerings on equipment from Nortel Networks, including Nortel's OPTera Packet Edge system, the BPS 2000 Ethernet switch, Passport 8600 Ethernet switch, and the Preside management system.

WorldCom's has attached this gear to its underlying SONET network using a prestandard implementation of RPR technology to provision and manage local nodes. So far, all services are point-to-point from the customer's perspective. WorldCom says it plans to offer switched services too, such as the switched Ethernet VPNs offered by BellSouth. They also say they're looking at alternative suppliers to Nortel.

Is WorldCom tiptoeing into a minefield? What's keeping WorldCom from miring itself in a sinkhole that exacerbates its already distressed financial situation—not to mention that its reputation is now in the tank and it's in Chapter 11 bankruptcy? Some industry experts say WorldCom will be okay. First, one of the biggest problems with Ethernet services is that they require high-priced fiber access on the part of business buyers. Startups don't have that access, but WorldCom certainly does. "The challenges faced by a Cogent or a Telseon are not the same for an RBOC or an established carrier," says Russ McGuire, Chief Strategy Officer at TeleChoice Inc. "The BLECs and ELECs must pay the incremental cost of getting into the customer's building. For others [like WorldCom and SBC], the cost is low." Even so, there are limitations. The enterprise-based Internet access and VPN services are available right now in selected cities (refer to Table 3-3), reflecting those locations where WorldCom has most access to buildings.

WorldCom also isn't up to speed with all its offerings. Although the intercity MAN links are supposed to reach 1 Gbps, they only support 622 Mbps (OC-12) right now. Enterprise-based VPN and Internet access is limited to 500 Mbps, although 1 Gbps speeds are promised. Spokespeople say the current shortfall is simply part of the planned rollout schedule. As demand builds, so will the available bandwidth.

Roderick Beck, an independent consultant based in New York, says WorldCom's new services offer a "user-friendly" interface for business customers who don't want to spend big bucks on a SONET interface to get higher speeds for corporate data. If these customers can get the bandwidth they need using their Ethernet interfaces, they'll probably go for it.

Beck expressed concern that Ethernet services could cut into the revenues that carriers get from private lines. That's an issue WorldCom claims it has under control. WorldCom will likely use "careful positioning" to avoid any revenue or services erosion.

Table 3-3

WorldCom Ethernet services profile

Service Name	Type of Service (ToS)	Customer Interfaces	Bandwidth Max	Location Offered	Price per Month
Metro and WAN private line	Corporate MAN links	Up to 1 Gbps	50, 150, and 622 Mbps	84 U.S. cities	Comparable to ATM and Frame Relay
Dedicated Internet	Enterprise access to Internet	10/100 Mbps and 1 Gbps	1 to 500 Mbps	Chicago, Dallas, New York, northern Virginia, San Francisco, San Jose, and Washington, D.C.	$1,200 to $200,000 per circuit
Enterprise private line/VPN	LAN-to-LAN and corporate network connections	10/100 Mbps	1 to 100 Mbps	Chicago, Dallas, New York, northern Virginia, San Francisco, an Jose, and Washington, D.C.	$630 to $20,000 per circuit

3.13.2.18 XO Communications XO Communications, formerly known as NextLink, offers GigE service as a LAN interconnection service to connect customers' LANs within metropolitan areas—effectively TLAN service. XO's service targets large companies with multiple locations and LANs in metro areas. The service is available in metro areas serviced by XO fiber, which currently includes 55 markets. XO also says it can provide nationwide availability based on customer demand. XO has about 200 route miles of fiber in each of its local markets. The service is available at Ethernet LAN port speeds of 10 Mbps, 100 Mbps, and 1 Gbps.

XO's Ethernet service does not require the customer to purchase any additional CPE. The service can be installed via most standard Ethernet LAN connectors (RJ-45 jacks), including 10BaseT, 100BaseTX, and 1000BaseLX. The customer must manage his or her own network and hand off to XO an IEEE-compliant 10BaseT, 100BaseTX, or 1000BaseLX Ethernet connection. XO is responsible for maintaining, troubleshooting, and monitoring the Ethernet transport equipment.

Key: To complement its optical Ethernet service, XO also deploys 100 Mbps fixed wireless Ethernet connections when necessary to connect customer's buildings that do not have access to optical fiber.

XO also plans to offer *virtual LAN* (VLAN) service and long-haul Ethernet using a combination of layer 2 and 3 technologies. XO offers its Ethernet services using Cisco equipment. The 10 and 100 Mbps Ethernet transport is over SONET, whereas 1 Gbps Ethernet transport is over DWDM.

3.13.2.18.1 DWDM Service In July 2001, XO announced the introduction of OC-12 (622 Mbps), OC-48 (2.5 Gbps), and OC-192 (10 Gbps) metro wavelength services in each of its 62 U.S. markets based on DWDM technology. The XO *Metro Wavelength Service* (MWS) is a protected service that provides carrier and enterprise customers with high bandwidth and protocol transparent flexibility at a low cost per megabit. By offering a protected service, XO can manage and maintain its customers' services and provision additional metro wavelengths, all without service interruption in a plug-and-play manner.

This is made possible via XO's deployment of DWDM equipment in its metro fiber networks. XO MWS provides every customer with a wavelength of light on a fiber-optic strand that is dedicated to that customer's service and can operate at different bit rates, and support multiple transport protocols (such as SONET, ATM, and IP).

Using its robust metro fiber networks, the XO MWS currently delivers up to 33 separate wavelengths of capacity per fiber strand, and provides a competitive advantage to customers by offering greater network capacity and scalability at lower transport and capital costs. XO launched its MWS to offer carrier and enterprise customers the ability to realize dedicated and protected OC-12 to OC-192 (622 Mbps to 10 Gbps) capacity between multiple sites within a metro area. "XO is one of the first broadband communications service providers to roll out wavelength services to the enterprise market across 62 U.S. markets," says Nick Maynard, an analyst with the Yankee Group. "XO's extensive metro fiber assets and long-haul capacity will play a fundamental role in allowing the company to gain market share in this space."

Key: XO is selling the wavelength services throughout its markets in the United States with a 1,100+ person direct sales force that include teams of data sales specialists and experienced national account sales executives.

Figure 3-9 delivers a graphical view of XO's nationwide network.

Figure 3-9
XO Communications'
nationwide network
backbone

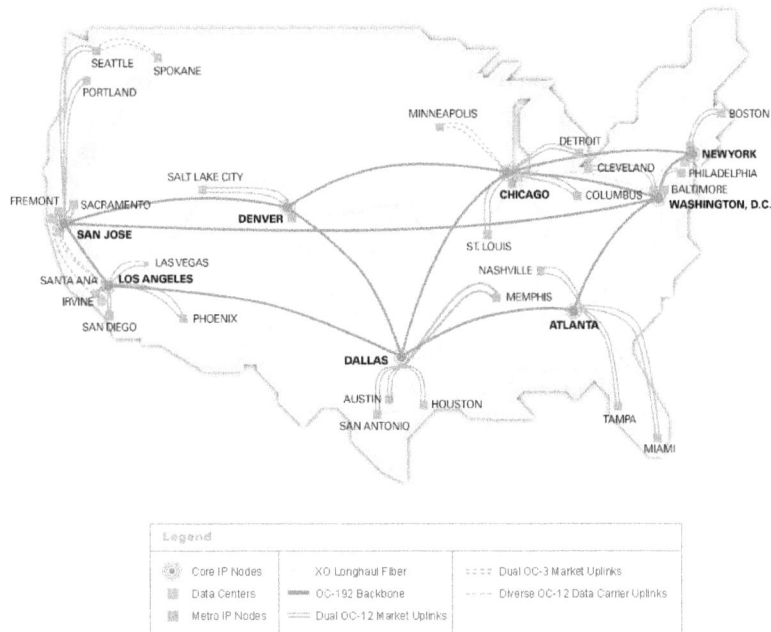

Unfortunately, XO Communications is one of the many telecom service providers who fell victim to the anemic economy in 2001 and 2002, and they declared Chapter 11 bankruptcy in June 2002. Nevertheless, it's likely the carrier will emerge somehow from bankruptcy, in part or in whole. Like WorldCom, some or all of its service offerings—including the ones listed below—will possibly survive in some form.

3.13.2.19 Yipes Communications Yipes Communications was one of the pioneers when it came to delivering super-fast connections to businesses using Ethernet as the last-mile connection to data or hosting centers. The San Francisco-based company was founded by a collection of former Nokia execs and Baby Bell defectors, and received $291 million in private funding.

Sadly, Yipes Communications filed for Chapter 11 bankruptcy protection in May 2002 after what appeared by all accounts to be a solid, trailblazing effort in the metro Ethernet space. Then in July 2002, Yipes began efforts to reemerge from bankruptcy amid accusations that the company filed for bankruptcy protection unnecessarily. The unfortunate part is that even if Yipes truly emerges from Chapter 11, the company will never be able to regain the trust of its former and potential customers. It's credibility is now tattered.

A history of Yipes is still included in this section for several reasons. First, it's important because Yipes truly was one of the first independent (non-ILEC) trailblazers in the metro Ethernet marketplace. Yipes was the first true ELEC. They were very aggressive, original in their approach, and one of the first companies to apparently convince the enterprise world that a pure-play Ethernet architecture could work, and work well. Second, Yipes' approach was watched carefully by the industry and won the praise of many analysts.

Yipes is a privately held network service provider based in the San Francisco Bay area, serving customers since 1997. Before co-founding Yipes, former Yipes CEO Jerry Parrick was best known for building US West's !NTERPRISE division into a $500 million business. In July 1999, Yipes obtained its first venture capital funding ($13.8 million), and in subsequent rounds of funding, it raised an additional $77 million and then $200 million. This produced a total of $291 million in private equity capital.

Yipes built an all-optical IP network based on Ethernet rather than incumbent telecommunication protocols (such as ATM and SONET). The Yipes network is comprised of 1,600 route miles of lit fiber and 32,000 lit strand miles, with an average of 8 strands per connection.

Key: Instead of offering a voice-oriented architecture of traditional networks, Yipes offers a native IP network that eliminates intercarrier conversions for on-net customers. Fewer points of conversion mean fewer opportunities for failure with greater reliability, the highest possible data integrity, and the lowest possible latency.

As of December 2001, Yipes served 21 markets across the United States. Customers included Fortune 1000 companies, real-estate companies, financial institutions, ISPs and ASPs, web-based businesses, software developers, law firms, major medical facilities, universities, school districts, and government agencies. Yipes focused its sales efforts on eliminating metro and regional DS-1 and DS-3 bottlenecks for enterprise customers (75 percent), but it also serves carrier customers (25 percent) in a wholesale manner.

Yipes was the defining provider of managed optical IP networks. They leveraged the elegance of native Ethernet technology to provide Just-In-Time Bandwidth[SM] that was smoothly and quickly scalable from 1 Mbps to 1 Gbps in 1 Mbps increments.

The company offered five main services across a logically meshed network, which means that these were point-to-point, point-to-multipoint, or multipoint-to-multipoint services. Yipes' very affordable services include the following:

- **Yipes MAN**   A LAN-to-LAN interconnect service (TLAN). Customers select speeds from 1 Mbps to 1 Gbps, and Yipes provides and manages routers on customer premises. The average bandwidth purchased by Yipes MAN customers was 35 to 45 Mbps. Yipes MAN is provided using layer 2 VLAN tags based on IEEE 802.1

- **Yipes WAN**   A LAN-to-LAN interconnection service that offers bandwidth from 1 to 100 Mbps to connect LANs in different cities that are part of Yipes MAN (that is, on net). The service is fully managed and utilizes MPLS at the network edge.

- **Yipes NET (high-speed Internet access)**   Offers Internet access from 1 Mbps to 1 Gbps. Yipes installs and manages routers at the customer premises to hand off Ethernet traffic to the customer's LANs. The average bandwidth purchased by Yipes NET customers is 16 Mbps.

- **Yipes WALL (managed firewall service)**   A scalable, wire-speed stateful firewall and security monitoring service designed and managed to complement the NET and WEB services.

▪ **Yipes TOUGH**   Redundancy services for Yipes MAN and Yipes *National Area Network* (NAN) customers.

▪ **Yipes NOW**   Bandwidth delivered to existing circuits overnight in 1 Mbps increments.

▪ **Yipes WEB (managed web services)**   Offers managed data services that include professional services, managed hosting, and collocation, as well as network design and implementation.

Yipes' Application-Aware Network^{SM} empowered customers to manage bandwidth and QoS on a real-time, application-specific basis. Supported applications varied, including streaming media, VPNs, options trading, and basic headquarters to branch office connectivity. Yipes' variety of services provided connectivity from 1 Mbps to 1 Gbps, including delivery and installation of access equipment at the customer's building for connecting a company's LAN to the Yipes network. Yipes provided the customer with a standard 10/100 BaseT Ethernet RJ-45 port at the enterprise, which is dedicated to that enterprise customer.

As the poster child for the ELEC market, Yipes attracted the most attention of the predominantly retail ELECs. Its footprint encompassed 107 rings across roughly 30,000 fiber-strand miles in 20 markets. Total customers exceeded 1,000 and included service providers (ISPs and ASPs), manufacturers, professional/technical services, publishing, healthcare, government, e-commerce, and utilities businesses.

In mid-2001, Yipes metropolitan networks were operational in 20 cities: Atlanta; Boston; Chicago; Fort Collins, Colorado; Dallas; Denver; Fort Lauderdale; Fort Longmont, Colorado; Houston; Miami; Palo Alto; Philadelphia; Pittsburgh; Riverside, California; San Diego; San Francisco; Santa Clara; Seattle; Washington, D.C.; and Worcester, Massachusetts.

Although Yipes initially targeted small businesses, the company later targeted service providers and large and medium-sized companies. Yipes acquired (leased) fiber from over a dozen different dark-fiber providers, including MFN, municipalities, and utilities. Yipes' long-haul IP backbone was provided by UUNet, Level 3, Savvis, Genuity, and Qwest.

According to Yipes, it could change bandwidth within three hours remotely from an NOC. The company rolled out a web-based interface in late 2001 to enable customers to initiate bandwidth changes themselves. Yipes priced its service on an ICB basis, factoring in the cost of building out fiber laterals to the customers' premises, so pricing was somewhat distance sensitive. Yipes services were fully managed and included the cost of CPE.

Yipes offered a standard SLA including network availability (99.99 percent), latency (less than 10 ms), and on-time installation. They deployed the following network management tools: HP OpenView *Network Management System* (NMS), Remedy trouble ticketing system, Micromuse Netcool (network alarms), Concord Network Health (traffic usage stats), and Jyra (latency measurements).

The Yipes network was built from the ground up, using diverse dark fiber (20-year IRU) in the metro area from about 15 or 16 different dark-fiber providers around the world. However, even with the large number of carriers providing the dark fiber, the Yipes network completely bypasses the LEC networks. At least two fiber providers are used in each metro market. Depending on the fiber source, physical diversity may be provided to the *building POP* (BPOP). Customers can also arrange for physically redundant paths if they are not readily available to a specific building.

3.13.2.19.1 SLAs and Network Management

Yipes has an SLA that comes standard with all service. For on-net service, the company aims to provide 100 percent network reliability and credits customers for one day of service for every hour of unscheduled downtime (up to one month's recurring charges). For WAN service, the availability guarantee applies end to end (including CPE and Yipes local ring), but for 99.7 percent network availability. The small print defines outages as those that occur for more than two hours at three or more sites. For on-net links, if the average monthly latency is more than 10 ms within a specific region, the customer is credited with 10 percent of one month's recurring charges. NAN latency guarantees are, however, distance sensitive. Latency guarantees haven't been established yet for off-net WAN service.

Yipes offered customers web-based network utilization and SLA reports, 24×7 toll-free access to live technical support, and 24×7 network monitoring and fault detection. Yipes provides a customer service portal where customers can access their network usage and performance statistics (including accumulated SLA credits) via the web.

3.13.2.19.2 Yipes Buys Broadband Office

In late 2001, Yipes bought the access rights to 3,622 buildings from *BroadBand Office* (BBO) for $2 million in cash, $2.5 million in equity, and $900,000 in a forgiven-debtor-in-possession bridge loan Yipes made to BBO. The provider's fiber passed within one-quarter mile of 240,000 buildings. Was this really a beneficial move? Acquiring a company whose holdings still required the construction of thousands of lateral fiber connections? If Yipes could not get to the building in areas where cities are less than cooperative about allowing providers to lay fiber, it intended to use T1 (1.5 Mbps) lines as a temporary solution.

This was easier as copper goes everywhere. Yipes emerged from bankruptcy as Yipes Enterprises, operating in only half of its original markets (10 cities).

Table 3-4 summarizes who the MAN players are, their services and target markets, and the type of equipment they all use.

Table 3-5 lists the key wholesalers of dark fiber in the metro space. These are the companies that the MAN providers lease services from in order to build their metro networks.

3.14   Carrier Partnerships

Some carriers realize that partnerships in the telecom game can be symbiotic, which is shown in the following section.

3.14.1   Telseon and Broadwing

Broadwing is using Telseon's network to provide GigE access connectivity to its customers. In this arrangement, Telseon acts more as a carrier's carrier, while Broadwing controls the customer account. The two networks complement each other well. As described by Bob Klessig, the vice president of Telseon, Telseon places a "service interface unit," essentially a Riverstone Ethernet router, at the customer location. The router takes Ethernet traffic from the CPE and transmits over fiber leased (wholesale) from Level 3 Communications to centralized Telseon facilities. It is collected and combined with other Telseon traffic by a GigE access switch, which is also manufactured by Riverstone. Now at GigE speed, traffic is handed off to a larger GigE metro core network that handles all of Telseon's traffic, again over leased fiber, in a given metro area. From there, traffic connects into Broadwing's metro optical network.

At Broadwing, higher-level optical switching begins. Once again traffic is aggregated, this time by large CIENA Corp DWDM transport systems. These systems are viewed as on ramps to Broadwing's long-haul optical core network. The grooming and transport process amounts to a fairly simple picture of optical Ethernet and DWDM, which is not very different from the IP/DWDM models already in use by some of the long-haul carriers. But it's just a start. This architecture could at long last be a successful business model for alternative local access.

Table 3-4 MAN service providers

Carrier	Target Market	Services They Sell	Equipment They Use
Cogent Communications	MTUs, enterprises, ISPs, ASPs, SSPs, and CLECs	100 to 1,000 Mb Ethernet	Ethernet switches and aggregators, DWDM transport, and multiprotocol transport
Giant Loop	Large enterprises, including financial institutions	LAN, WAN, VPN, and managed optical network services using GigE, SONET, and Fibre Channel protocols	Multiprotocol transport platform
Looking Glass Networks	IXCs, ILECs, CLECs, and ISPs	LAN, WAN, VPN, storage networking, interfacility DWDM, and SONET wavelengths	Optical switch, DWDM transport, and SONET transport
Sphera Optical Networks	IXCs, ILECs, CLECs, ISPs, and large enterprises	Interfacility connectivity and dedicated access transport via DWDM wavelengths	DWDM transport
Telseon	Long-haul and regional service providers, ISPs, ASPs who are looking for a GigE service component	Last-mile GigE connectivity	Ethernet routers, switches, and aggregators
TWT	Large and medium-sized enterprises andMTUs	Dial tone, ATM, LAN, WAN, VPN, hosting, videoconferencing, and dedicated access transport	ATM switches, DWDM transport, SONET transport, softswitch, and Class 5 (end-office) switch
XO Communications	Large and medium-sized enterprises and MTUs	Dial tone, ATM, LAN, WAN, VPN, hosting, videoconferencing, and dedicated access transport	ATM switches, DWDM transport, SONET transport, softswitch, and Class 5 switch

Source: America's Network "GigE Evolution," Steven Titch, January 15, 2002.

Table 3-5 Dark-fiber wholesalers

Service Provider	Target Market	Services They Sell	Equipment They Use
AFS	CLECs, ILECs, IXCs, and ISPs	Dark-fiber capacity and wavelength management	Optical fiber
Level 3 Communications	IXCs, ILECs, CLECs, ISP, and large enterprises	IP, ATM dedicated Internet, LAN, WAN, VPN, *voice over IP* (VoIP), dark-fiber capacity, and wavelength management	Softswitch and optical fiber
MFN	CLECs, ILECs, IXCs, ISPs, and ASPs	Dark-fiber capacity and wavelength management	DWDM transport, Ethernet switch, and optical fiber

3.14.2   All Others

The entire metro Ethernet arena is a very incestuous marketplace. Almost all of the ELECs and BLECs relied on MFN for dark fiber to build their metro networks until MFN declared Chapter 11 bankruptcy and ruined their reputation. Many other metro carriers also relied heavily on Level 3, Qwest Communications, and other dark-fiber providers in order to build their networks. Soon, it's possible that the independent metro carriers will even start to rely on the ILECs for last-feet connectivity to their target customers. This is even more likely if more wholesale (fiber) carriers continue to declare bankruptcy.

3.15   Pricing Strategies

According to a model presented by consultant Michael Kennedy at a Massachusetts Telecommunications Council meeting in early 2002, the RBOCs can compete on price with startup metro ESPs and still make a profit. This means the RBOCs will start offering more metro Ethernet services and users will be able to procure metro area Ethernet services from their traditional carriers.

Ethernet services from companies such as Yipes, Cogent, and Telseon are generally priced to undercut RBOC data service prices. For example, Cogent offers 100 Mbps Ethernet service for $1,000 a month. The act of offering drastically lower prices than its competitors is known as disruptive pricing. By contrast, a 45 Mbps DS-3 service could cost around $6,000 per month, according to Kennedy.

Users can also increase the bandwidth of their Ethernet services in very small increments, typically 1 Mbps. However, private lines must be purchased one T1 at a time, and after six to eight have been bought, it's cheaper to buy a DS-3 circuit. This means that users may end up paying for bandwidth they don't use because they're forced to make such a huge jump in bandwidth. This is due to the historical lack of an available service offering between DS-1 and DS-3 services. For decades, enterprises have not had a feasible way to increase bandwidth in small increments. The jump from DS-1 service to DS-3 service is big (44 Mb) and has a price increase to match.

Kennedy has claimed, however, that the RBOCs can price their Ethernet services so that they don't compete with their current T1 business, but still attract customers.

The RBOCs could price the incremental Ethernet services (1 Mb at a time) so that customers end up paying more for a 45 Mbps Ethernet service than they would pay for a DS-3. According to Kennedy, customers will pay that premium because most would rather pay for service in increments versus migrating to a DS-3, which would require new wiring and equipment.

3.15.1   Cost Analysis

Startup service providers are positioning Ethernet services as low-cost alternatives to leased-line offerings provided by incumbent service providers, primarily T1 leased lines. Average pricing for a T1 is $700 to $1,200 a month (for 1.544 Mb of bandwidth), depending on the distance between the endpoints. ESP Cogent, by contrast, promotes its 100 Mbps service at $1,000 a month. Comparing Cogent pricing versus T1 pricing on a per megabit basis yields the following result:

Cogent: $10 per megabit

Incumbent T1 price: $453 to $777 per megabit

So Cogent's Ethernet service pricing is at least 45 times less than incumbent T1 pricing. Ethernet service pricing per megabit is generally much lower than typical T1 offerings. Telseon's service is priced 16 times lower than incumbent T1 pricing. Cost is a key driver in end-user migration from TDM services to Ethernet services based on optical Ethernet infrastructure.

Looking at equipment costs, the argument is similarly strong in favor of Ethernet offerings over traditional SONET products. A gigabit port on an Ethernet switch list currently costs about $1,200, or $1.20 per megabit. By contrast, a SONET OC-12 ADM lists at about $35,000, or $56 per megabit. Therefore, GigE equipment costs are 47 times lower than SONET equipment costs.

The hardware savings argument also resonates with the service provider's customers, who must buy PoS interfaces or Ethernet interfaces for their routers. To hook up to an OC-3 SONET service, the customer must buy two PoS interfaces for about $10,000 each—one for each end of the connection. To receive comparable bandwidth with Ethernet service, the customer needs two Fast Ethernet interfaces (100 Mbps) for $400 each. It's hard to argue the cost effectiveness of Ethernet given these facts.

3.15.2 Pricing

Pricing for optical Ethernet services varies by service provider. Prices can range from $1,570 to $4,040 per month for 100 Mbps services from Bell-South and Telseon, respectively. At GigE rates, SBC's tariff is $5,000 to $7,000 depending on interoffice mileage requirements. XO's tariff is $14,000. A number of service providers who offer or plan to offer optical Ethernet service use the following planning numbers for customers' business cases: $1,100 a month for 10 Mbps; $2,300 a month for 100 Mbps; and $11,000 a month for 1 Gbps.

Key: On a cost/bit basis, these prices are 30 percent of the price for T1, 22 percent of a DS-3, and 20 percent of an OC-3.

3.16   Metro Area Ethernet Equipment Vendors

Now for the equipment end of the puzzle. Just like the service provider segment, in the equipment segment many new manufacturers have sprung up in the last few years in an effort to take the battle to the incumbent hardware companies such as Cisco, Lucent, and Nortel. Some of these companies are gutsy and very innovative (Extreme Networks). Some of them are very efficient, and make scalable, high-quality equipment (LuxN). Some of them are making serious inroads in the marketplace and taking decent market share away from the incumbents (Riverstone Networks, Foundry, and Atrica). An equal number of incumbents and startups are floundering. Lucent and Nortel especially seem to be struggling just like many of the new players. Here's a look at the players and their products.

3.16.1   The Vendor Marketplace

Just as the metro Ethernet space has attracted many new service providers who are using new, somewhat narrowly defined business models, the equipment vendor space that supports this effort has also seen many upstarts enter the game since 1999. Many of these companies will not exist in their present form by 2005. This will be due to acquisitions or mergers with other carriers, Chapter 11 bankruptcy, or Chapter 7 bankruptcy. This market has too many players for all of them to survive and thrive. As in nature, only the strong will survive. Many firms will fold due to one or more of the following pitfalls: inept management, lack of venture capital funding, slim profit margins, revenue streams flowing slower than expected, poor marketing, lack of talent, intense competitive pressures, or time to market. Nevertheless, the metro presents a very attractive opportunity for equipment manufacturers that successfully address the burgeoning market in terms of appropriate functionality, pricing, time to market, and price points to support evolving network architectures. However, driven by uncertainty in network traffic, architecture, and actions of incumbents, significant risks will still remain for investors and equipment providers.

The equipment needs of metro service providers differ substantially from the long-haul market. More emphasis was placed on lower port costs and equipment functionality (that is, capacity and space density). Other important criteria include scalability, upgradability, and support of non-SONET interfaces as well as traditional SONET.

Two distinct customer segments with diverse needs have emerged in the metro (GigE) space: service providers (carriers) and enterprises (businesses). Incumbent equipment vendors such as Cisco, Nortel, and Lucent are working in this market, along with new entrants. Both types of vendors have aligned their business models against these two segments: metro transport players and metro access players. Some service providers are operating in both segments (such as the ILECs).

Key: A key driver for the future success of equipment vendors will be how well they integrate emerging component technologies and operating system software into their product lines.

In the metro access segment, GigE is expected to emerge as the primary local access packet data protocol. Differentiating products should support low-cost data transport, along with newer applications such as VoIP and IP VPNs.

The metro space has significant opportunity and risk. Despite the strong projected growth in metro bandwidth, many real uncertainties exist as to which of the new service providers will succeed. Next-generation equipment performance also presents a source of uncertainty, as much of it is just beginning carrier testing and is not yet standardized.

3.16.2 Common System Requirements

Every networking environment is unique, but all networks have common denominators. Enterprises and carriers alike share a need for high degrees of protocol support, scalability, reliability, interoperability, ease of installation and management, management functionality, and system efficiency in their optical-networking platforms.

3.16.2.1 Protocol Support Systems should support all data-centric application traffic up to 2.488 Mbps (OC-48) to include the following protocols: ATM, ESCON, 10 Mb Ethernet, Fast Ethernet, *Fiber Distributed Data Interface* (FDDI), Fibre Channel, *Fiber Connectivity* (FICON), GigE, SONET, *Symmetrix Remote Data Facility* (SRDF), DS-3, and 10 GigE. Obviously, protocol transparency is optimal.

FICON is a high-speed *input/output* (I/O) interface for mainframe computer connections to storage devices. As part of IBM's S/390 server, FICON channels increase I/O capacity through the combination of a new architecture and faster physical link rates to make them up to eight times as efficient as *Enterprise System Connection* (ESCON), IBM's previous fiber-optic channel standard.

SRDF provides business continuity in the event of both planned and unplanned outages. SRDF is an online, host-independent, mirrored data solution that duplicates production site data on one or more physically separate target Symmetrix systems.

3.16.2.2 Scalability of Optical Platforms Optical-networking solutions that have been developed from the ground up for metro enterprise transport support pay-as-you-grow expansion. The solutions enable customers to run only a few or up to 32 optical wave channels of native-speed traffic and evolve their network topologies—point to point, star, and rings —with the same platform. Furthermore, production traffic is not interrupted by the installation of additional hardware or software upgrades. Additional optical channels can be added as necessary, and capacity can be increased at each network node using add/drop capabilities.

3.16.2.3 Reliability Enterprises and carriers demand line protection for point-to-point links as well as full path protection for ring structures. In either case, service restoration of less than 50 ms is of vital importance. Otherwise, the performance of latency-sensitive applications such as VoIP and video could be jeopardized.

Additionally, all system components must be redundant with multiple failover scenarios in mind. For example, enterprises and carriers frequently seek optical platforms with redundant power supplies and hot-standby functionality.

3.16.2.4 Interoperability (Openness) and Network Management
Standards compliance can be a make-or-break factor in buying decisions in any area of networking. Support for three key network management protocols ensures that an optical-networking platform can be easily integrated with most existing enterprise and carrier infrastructures. These three protocols are

- *Signaling Network Management Protocol* **(SNMP)** The simple network management language most widely used by enterprises and emerging carriers
- **TL-1** The management interface used in most North American incumbent carrier environments
- **Q3** Another management protocol used extensively in Europe and gaining prominence globally

Enterprises and carriers choosing an optical platform are also interested in standards regarding *single-mode fiber* (SMF) (*International Telecommunication Union Technology Sector* [ITU-T] G.652, G.653, G.654, and G.655), electromagnetic compatibility, electrostatic discharge immunity tests, immunity to conducted disturbances induced by RF fields, mechanical and electrical safety, and optical safety.

Enterprises and carriers require sophisticated management functionality such as configuration management, service provision and activation, service testing, and fault monitoring. For example, in the area of system monitoring, some carriers require specially designed clocks that can be remotely controlled. This is so network operators can provide distinguished tariff speed for each customer and service according to service value (rather than flat rate). Carriers must be able to monitor, control, or even configure bit rates via the management system.

Because today's leading optical vendor platforms rely on the network-management protocols so prevalent in both enterprise and carrier environments (SNMP, TL-1, and Q3), they are user friendly. Network administrators can become comfortable using the systems with no more than a week of training. Some vendors' products support plug-and-play installation, a major benefit to enterprises and carriers whose network support staffs are strapped for time. Platforms that aren't plug-and-play require complex software configuration during installation.

3.16.2.5 System Efficiency Like ease of installation, system size is a major concern and it varies greatly among the various platforms. Many available products with enterprise listed among their target applications were initially designed for carrier deployment. Because of that, they are too big. Some systems even require their own dedicated refrigeration units! Enterprises want platforms that fit standard 7-foot, 19-inch networking cabinets. A fully loaded chassis should not exceed 25 kilograms or 150 watts in power consumption.

3.17 Market Forecast

Table 3-6 shows a forecast for metro optical equipment revenue for 2001 to 2005 by technology segment. Note that the decline of legacy SONET equipment is forecasted to start in 2004. The technologies listed in this table will be reviewed in detail in Chapter 4, "Competing Technologies," and Chapter 5, "Complementary Technologies and Protocols."

The challenge for the upstart Ethernet switch vendors is to move from targeting the remaining CLECs to targeting the traditional ILECs whose buying power dwarfs that of the CLEC community. Ethernet upstart equipment vendors will likely partner with an established equipment supplier or, depending on their long-term financial viability, will look to be acquired by these mature players. This approach would help to ensure their long-term viability. For example, popular upstart Riverstone Networks has attempted to break into the incumbent market through an agreement with incumbent favorite Tellabs.

Table 3-6

Metro optical equipment revenue by technology segment ($ billions)

	2001	2002	2003	2004	2005
Legacy SONET/ *Synchronous Digital Hierarchy* (SDH)	$10.0	$10.4	$10.0	$9.0	$8.0
Next-generation SONET/SDH	$1.0	$1.3	$2.3	$3.8	$5.2
DWDM	$1.5	$2.0	$3.1	$5.0	$6.3
Optical Ethernet	$.05	$0.7	$1.4	$2.3	$4.1
Total	$13.0	$14.4	$16.8	$20.1	$23.6

Source: Cahners In-Stat/MDR (2002)

3.18   GigE Equipment Development

Three stages of development for GigE switch vendors have been identified:

1. **Enterprise or hosting centers**   Functionality in this market segment includes basic layer 3 switching and QoS features (such as multiple traffic queues per port, with prioritization based on layer 1 to 4 information) and basic VLAN capabilities.

2. **Platforms for single-MAN TLAN service**   These are high-end, scalable enterprise hosting switches with functionality that makes the product appropriate for use as a service operated by a city, for example, versus a carrier service. Ideal deployments include cases where the network must scale to support much higher numbers of customers and more diverse customer requirements, and interoperate with WAN services. Functionality includes higher port density (more ports per box/rack and higher bandwidth rates per port), fault-tolerant (redundant) power supplies, ToS and DiffServ QoS, and multicast support. ToS (a field in IP packets) and Diffserv prioritize and manage traffic according to traffic type—for example, voice, video, or data.

3. **True carrier-class products**   Carrier-class features would include the following:

   - *Network Equipment Building Systems* (NEBS) compliance level 3
   - ATM and PoS interfaces for legacy customers
   - QoS that includes latency guarantees
   - Support for *Border Gateway Protocol 4* (BGP4) and MPLS
   - VLANs that support *customer views*, meaning customers can see traffic and management statistics related to each VLAN
   - Enhanced security
   - Extended gigabit distances
   - Rate limiting by port and protocol
   - Trunk aggregation (standard 802.3ad)
   - Easy integration with existing carrier element and network management systems (EMS and NMS)
   - Enhanced billing features
   - 10 GigE capability

More than 90 percent of incumbent GigE switch vendors are in the first or second stage of equipment development to meet these requirements. A number of other vendors, including Cisco, Nortel, and Foundry, have attributes from both the second and third categories. Only Riverstone and Extreme are currently offering true carrier-class products based on the criteria listed previously.

Key: The GigE switch vendor segment, like virtually every product segment in the telecommunications and data communications market, is packed with startups that outnumber established vendors by a ratio of at least 8:1.

In the metro Ethernet market, they have gravitated to one of three differentiating attributes:

- **Scale and density**   The Canadian startup VIP Switch promised 640 GigE ports per rack. The V-MAN 160G scales by adding up to 16 chassis, which can be managed as a single logical entity. A full systems configuration scales to 2,560 GigE ports in 4 racks. Obviously, this product would be best suited for the function of aggregation.
- **SONET-like protection**   Dynarc, Corrigent, Luminous Networks, and Lantern Communications are promising below 50 ms restoration based on RPR.
- **Flexible access**   Appian Communications and World Wide Packets are developing last-mile solutions based on Ethernet. Although many of the Ethernet vendors are focusing on access alone, World Wide Packets has a unique architecture that includes both carrier- and enterprise-class products.

Appian Communications is developing an *Ethernet-over-SONET* (EoS) solution based on ITU X.86 standards (see Chapter 5). Although Appian delivers only one platform at this time—the *Optical Service Activation Platform* (OSAP)—it can be positioned at multiple points in the network:

- At the optical service edge, such as in a building basement (MTU)
- An end office (Class 5 CO)
- A carrier hotel
- A collocation facility

Thus, it provides a measure of flexibility as well as uniformity, which reduces sparing costs.

3.18.1 Progress in the 10 GigE Equipment Arena

Sycamore and CIENA are maximizing the number of GigE streams that can be packed into OC-48 and OC-192 wavelengths. 10 GigE ports will become the interface of choice for interconnection as 10 GigE use reaches the economies of scale that have traditionally characterized Ethernet and Ethernet's annual declining cost curve of 30 percent. Carriers and service providers recognize that 10 GigE interconnection is less expensive to provide than other alternatives.

Key: Although shipments are not expected in quantity until late 2002 or early 2003, initial price points per discussions with vendors for the 10 GigE WAN standard (WAN PHY) range from $10,000 to $50,000 per port. In comparison, a 10 GigE PoS interface lists for $295,000. As the declining cost of components, manufacturing, and research and development drive down systems costs, Ethernet (and GigE in particular) will move further out into the network: from core to edge to access.

The argument in favor of Ethernet services and Ethernet equipment focuses largely around cost-to-performance ratio. Cost will continue to be a key driver in end-user migration from TDM services to Ethernet services, based on optical Ethernet infrastructure.

Cisco, Tellabs, and 3Com are a few vendors that are specifically pinning high hopes on the metro optical market for new revenue growth. To bolster their metro optical product lines, many vendors have acquired startups with promising technology.

3.19 Multiservice Provisioning Platforms (MSPPs)

Equipment manufacturers are trying to attain economies of scale with the development and deployment of new platforms known as MSPPs. These boxes portend to manage the traffic of multiple disparate protocols all on one box. There are advantages and disadvantages to these boxes, as noted here.

3.19.1 Multiprotocol Provisioning

MSPPs are a newer-generation all-purpose platform, sometimes also known as *god boxes*. MSPPs offer a mixture of TDM, IP, Ethernet, and ATM services. They can be scalable, which is usually accomplished through DWDM that can be integrated within or provided by another *network element* (NE). Each vendor may have two or more major MSPP-type products —one for edge services and one for the metro core—but a single box addresses legacy, SONET ADM requirements with front-end cards for advanced protocols.

Key: One of the key design features of MSPPs is the choice of either multiple switch fabrics for each protocol transported, or a single switch fabric for any and all protocols being considered.

Although many hardware vendors are competing for this market (Cisco and Nortel being the dominant players), the Yankee Group believes other vendors are worth watching. It's believed that these vendors are either leaders in their categories or the offerings exhibit most of the features that characterize their genre.

Startup carriers know that their market advantage comes from their lack of legacy networks and their exclusive focus on data. It's not that the incumbents don't see the opportunity. Verizon Communications, SBC Communications, Qwest Communications, and BellSouth all jumped at the chance to become charter members of the MEF, which was founded in July 2001.

The problem is that the RBOCs have to deal with the legacy of SONET transport platforms in their metro networks and a reliance on voice traffic for revenue. They also have many customers using ATM, frame relay, pure IP, and DSL. The legacy problem is not isolated to the ILECs. Many of the established broadband and data CLECs, such as Covad Communications, TWT, XO Communications, and the local access units of AT&T and Sprint face legacy issues as well. One solution, which is only just emerging, involves the deployment of versatile boxes that can aggregate not only Ethernet, but other protocols as well onto high-speed optical transmission systems. These versatile boxes are MSPPs, which can groom traffic directly to OC-48 or OC-192 speeds and then route them to a SONET or DWDM system.

Key: Although a lot of promises are being made, especially pertaining to next-generation SONET, few equipment vendors developing MSPPs have products out of the lab or beta test stage as of early 2002.

Along with startups such as Atoga Systems, Native Networks, and White Rock Networks, many of the current SONET players, including Tellabs, Alcatel, and Nortel Networks have announced plans for MSPP products. The success of any vendor in this space hinges on how well they can migrate and manage the quality levels of merged voice and data services. The challenge is augmented because IP—the de facto standard for data transmission—contains no inherent QoS characteristics at this time.

"To carriers, the issue is whether to go with MSPPs that offer economies of scale, but risk creating a configuration mess at the same time," according to Tracy Vanik, a senior analyst at RHK. "The alternative is to go with a number of different access switches, each handling a different protocol. It costs less in terms of management, says Vanik, but could end up costing more in terms of provisioning time. Any decision comes down to the service provider's overall strategy on data services." From Vanik's perspective, a service provider must decide exactly how they want to approach the market for high-capacity bandwidth provisioning before they begin to make infrastructure decisions. This underscores the ILECs' central dilemma: They are struggling to determine the best way to replace innate voice revenues with data service revenues while simultaneously leveraging their legacy networks. It's not a simple problem.

But the ILECs still have options. Realistic and economically viable methods of integrating legacy TDM-SONET and IP-Ethernet-DWDM systems are available. The GigE players won't begin to make a noticeable dent in ILEC data revenues until 2004 or 2005. However, the longer the ILECs delay, the more influence they lose over the future direction of Ethernet-based network architecture and management design.

Key: The equipment needs of metro service providers differ substantially from the long-haul market, with greater emphasis on lower port costs and equipment functionality (that is, high capacity and low space density). This is because by its very nature, the metro market has a radically higher ratio of connectivity requirements than the long-haul market. This requirement is driven by the much greater number of subscribers and carriers that exist in metro markets.

Other important criteria for metro Ethernet equipment include scalability, upgradability, and support of non-SONET interfaces as well as traditional SONET. Another key driver should also be how well vendors integrate emerging component technologies and operating system software into their product lines. Newer products should also support traditional data transport (such as TDM-based private-line services) as well as newer applications like VoIP and IP VPNs. Offering QoS and high availability at the same time is also critical.

Next-generation equipment performance also introduces a source of uncertainty because much of it is just beginning carrier testing and is not standardized. Newer, disruptive technologies that are emerging may overshadow the cost advantages of current solutions.

Although god box systems that combine multiple transport and switching functions onto a single platform make economic sense, in operation they may be less than ideal. Telco operations and maintenance staffs are typically segmented into transport and switching functions, with no cross training between the two groups. For this reason, multifunction systems will be difficult to maintain when deployed in telco networks. If telcos eventually transform their internal personnel assignments—which they should to some degree—this issue will be minimized.

3.19.2   MSPP or Next-Generation SONET? Which Is It?

It is necessary to clarify how some emerging platforms are defined. A debate has been taking place over the labeling of vendor's newer generations of optical equipment.

The term *MSPP* is new, defining boxes that can manage the transport of both legacy protocols and newer disruptive technologies such as Ethernet. At the same time, many equipment vendors are introducing platforms known as *next-generation SONET* systems. There's considerable confusion in the industry regarding when a system is defined as one or the other: an MSPP or a next-generation SONET box.

Some call a new product category next-generation SONET; others call it an MSPP. Some industry analysts feel that neither term has been clearly defined, and that vendors are making up their own definitions and then claiming to be market leaders (how surprising).

MSPPs allegedly combine transport, switching, and routing platforms into an integrated system that enable service providers to offer bundled services flexibly and at a lower cost. By incorporating many service interfaces into one box, MSPPs supposedly eliminate the need for extra devices to deliver intelligent optical services. Vendors also claim that MSPPs improve SONET's efficiency by transporting multiservice traffic.

However, next-generation SONET platforms are touted to do much the same thing, depending on whom you ask. IDC research analyst Sterling Perrin says MSPPs "grew out of" the next-generation SONET market. "The MSPP category is really an umbrella category that includes next-generation SONET," said Perrin. "The lines are very blurry between the two, and because of that, there's a lot of confusion within both spaces." Perrin states also that original next-generation SONET boxes were single-platform SONET. They were all TDMs with an integrated cross-connect capability and EoS capabilities were made possible through a dedicated interface. "Then MSPP startups started emerging and taking this model and throwing all kinds of other functions into it, like layer 2 switching and layer 3 routing," stated Perrin. "That's when the question originated of where to draw the line between an MSPP and next-generation SONET system came up."

Marian Stasney, a senior analyst with the Yankee Group, says a box has to meet a couple criteria in order to qualify as an MSPP. "An MSPP box has to have a separate circuit (and module) for each service and a switching fabric for each type of traffic," she says. "A next-generation SONET box operates strictly at layer 1. It has an ADM function, optical cross-connects, and a DWDM backplane. On the other hand, MSPPs include switching and perhaps routing functions."

One vendor, Astral Point, believes that MSPPs and next-generation SONET systems are two distinct product categories. Nortel also views next-generation SONET and MSPP as separate and distinct markets. "Next-generation SONET works with the SONET speeds that are out there and drives them into a single, cost-effective aggregation point," says Joe Padgett, Nortel Director of Marketing, optical markets. "What we're seeing are these god boxes in MSPP. They're not granular, they don't support higher speeds, and they don't offer a smaller footprint."

Cisco, on the other hand, says the two terms define the same category, but also says the categories are a function of the definitions used. It claims that its *optical network system* (ONS) 15454 metro optical system qualifies as both an MSPP and a next-generation SONET platform. "There's an industry trend that's moving from SONET to multiservice and then embracing wavelengths in metro DWDM," says Rob Koslowsky, Cisco Director of Marketing, optical transport. "The driver: to embrace existing TDM services and handle higher bandwidth needs, but also to transition from voice-centric to data-centric transport systems."

Appian Communications says a subtle difference exists between MSPPs and next-generation SONET devices. MSPPs tend to support packet services more efficiently than next-generation SONET platforms. Next-generation SONET equipment has more port density, integrated ADM, grooming, and cross-connect capabilities predominantly for TDM, according to Appian.

Another optical access equipment vendor, Coriolis, seems to agree with Appian. Coriolis believes that MSPPs typically integrate packet switching into their platforms and can handle ATM, Ethernet, and frame relay circuits. Coriolis' OptiFlow system is apparently one such platform because it integrates a packet-switching fabric and ATM-switching fabric in the same box. Next-generation SONET equipment simply improves TDM density, according to Greg Wortman, Coriolis' vice president of marketing. "The Cisco ONS 15454 is the first of the next-generation SONET boxes," Wortman says. "Both Cisco and CIENA plug packet modules into their chassis. They don't actually integrate that service into their platforms. That's why neither is an MSPP."

CIENA's next-generation SONET box, the MetroDirector, integrates grooming and switching elements, and "provides basic SONET/TDM functionality in a higher-density rack-space and a lower cost while also providing support for GigE and ATM."

Grier Hansen, an optical and carrier infrastructure analyst for Current Analysis, sums the ambiguity up best. "There is a lot of overlap," he says. "That's why this has been such a problem. This is an impossible story to write, but one that needs to be written to help comb through the confusion."

The bottom line is that vendors will define and label a box to insert themselves into the market segment of their choice—where they think their greatest chance of success lies.

3.19.3 Optical Ethernet Provisioning Platforms (OEPPs)

Vendors building carrier-class GigE switches (layer 2/3 products) include Extreme Networks, Foundry, Nortel (Passport 8600), Atrica, Appian Communications, and World Wide Packets.

Key: Carrier-class switch vendors aim to bring as much resiliency to their products as possible without sacrificing the critical cost advantage that Ethernet holds over traditional SONET. All Ethernet switch vendors intend to price their products along the same cost curve that LAN-based Ethernet has followed.

The critical functionality required by carriers includes scalability, hardware redundancy, traffic-rate-shaping capabilities, CoS capabilities, and rapid restoration times. The current spanning tree restoration speeds (from 5 seconds up to 30 seconds) used by *optical Ethernet provisioning platform* (OEPP) vendors are inadequate, but the new rapid spanning tree algorithm promises restoration on the order of one second (see Chapter 5). Nortel is offering a restoration algorithm called *Split-MLT* for the Passport 8600 that also enables one-second restoration times.

One-second restoration appears to meet carrier needs for data and possibly for packetized voice traffic as well (VoIP). Although ESPs are betting on data in the near term, many plan to deploy some level of voice applications as well. Other carriers have used SONET-based products with Ethernet capabilities, which are defined by many in the industry as *SONET MSPPs*.

Most optical equipment, including metropolitan DWDM boxes and MSPPs, forms the infrastructure over which many types of services run, including TDM, storage protocols, and IP traffic. However, OEPPs are tied to a single ToS: Ethernet service.

Fortunately for equipment providers, Ethernet provides many benefits to service providers and their end customers. As indicated earlier, the argument for Ethernet services and Ethernet equipment centers largely around the cost-to-performance ratio (cost per megabit) based on the price per port and port density per chassis.

IDC believes that the market for OEPPs from 2002 to 2006 will consist primarily of greenfield players (new companies) focused on data services. Over time, OEPPs may gain traction with incumbent carriers as true replacements for their SONET infrastructure, but IDC doesn't believe this will happen on a large scale until at least sometime after 2005.

3.19.4   Next-Generation MAN Equipment Solutions: Three Classes

Many products are currently being touted as the right solution for the needs of the next-generation MAN. These products can be classified into three categories:

- SONET MSPPs
- Layer 2/3 packet switches
- Metro DWDM platforms

3.19.4.1 SONET MSPPs The legacy of the voice-optimized network and SONET-based equipment continues to live on, even in many new-generation products supposedly optimized for an IP-centric world. Many of these new technologies are based on new mappings of circuit-switched payloads into SONET frames. These products—SONET MSPPs—map traffic into different circuit pipes such as SONET STS-ns.

Key: A few of these next-generation SONET/circuit products support layer 2 and 3 packet-switching functions, but implement them as additional layers on top of the SONET/circuit layers. In this sense, they are just bolting switches and routers on top of SONET ADMs—even if those ADMs use nonstandard SONET payload mappings.

Although these solutions typically maintain most of the advantages of traditional SONET equipment (such as the ability to provide high-quality circuit services and sophisticated management, and the ability to be configured in simple, survivable ring topologies), they also do not completely eliminate its disadvantages. For example, such solutions cannot guarantee optimal bandwidth utilization for voice and data services since unused bandwidth in a given STS-n pipe is simply wasted. No statistical multiplexing function like ATM is available in this situation. By bolting a router or layer 2 switch on top of a SONET ADM, the cost and complexity associated with these multiple technologies is not eliminated. Finally, and most importantly, the presence of multiple technology layers does nothing to eliminate the necessity of managing these layers. This makes it difficult for carriers to scale their multiservice networks.

3.19.4.2 Layer 2/3 Packet Switches A second class of new MAN transport products uses native packet-switched solutions built for the LAN based on layer 2 switches or layer 3 routers. This is accomplished by connecting point-to-point links running Ethernet, PoS, or other protocols. Some of these solutions—especially the GigE switches—are capable of efficient bandwidth utilization and provide relatively sophisticated handling of differentiated data services. However, these approaches fail to provide the QoS and robustness required to transport real-time, mission-critical services. They don't guarantee end-to-end packet delivery with well-controlled latency, jitter, bandwidth, and packet-loss parameters. In particular, pure GigE solutions do not offer carriers the option of providing both data and high-quality circuit services (T1/E1 and DS-3/E3) on a single platform, which then forces the deployment of multiple boxes to accommodate customer needs. This would obviously defeat the purpose of service aggregation.

Key: Pure GigE approaches do not adapt well to the ring topologies that now dominate the metro network and do not exhibit the kind of network restoration times (50 ms or less) demanded for carrier-class transport equipment. This drawback will be eliminated when the RPR standard is released by 2003. This will coincide with GigE's rapid ascent in both the enterprise and carrier networks.

3.19.4.3 Metro DWDM   Another group of vendors is trying to adapt backbone optical transport technologies to the MAN. Metro DWDM platforms provide a significant increase in ring capacity based on their ability to multiplex 32 or more signals over a single fiber by mapping each signal to a different optical wavelength (frequency or lambda). Most of these platforms provide some form of basic ring protection switching and can transport a variety of traffic types in a protocol-independent manner. However, because these metro DWDM platforms typically map a given input stream onto its own wavelength, they do not provide a means for efficient, flexible aggregation of lower-rate streams onto the expensive DWDM facilities. Optical connectivity at the wavelength level also cannot be dynamically provisioned because it's based on fixed wavelength transmitters and multiplexers/demultiplexers.

Key: Some second-generation metro DWDM platform vendors are beginning to bolt SONET ADM (or cross-connect functionality) on top of the WDM layer to achieve some aggregation and dynamic provisioning functionality. This is achieved in the form of multiple SONET rings, with each ring using one wavelength. But in doing so, the vendors sacrifice protocol independence and incur all the costs, inefficiency, and inflexibility of standard SONET platforms.

Although the metro equipment space today is relatively small compared to the long-haul market, it represents a large growth opportunity. Equipment vendors that are leaders in important new technologies (such as metro DWDM, next-generation SONET, Ethernet, and VoIP) could be especially well placed. Given the likely shakeout in the metro service provider space and the considerable advantages that incumbents possess, new entrant backers certainly face risks. Additionally, significant uncertainty exists with respect to the carrier-scale operability and sustainability of the advantages created through disruptive technologies, such as emerging SONET-lite, metro DWDM, and GigE equipment.

Finally, the metro space presents a very high growth market for equipment vendors that meet the special needs of metro service providers by delivering radical cost-performance improvements or enabling new services. As a matter of fact, most of the startup service providers and incumbent carriers have already budgeted 10 to 25 percent annual equipment price declines in their business plans.

The equipment needs of metro service providers are very different from the equipment needs of long-haul carriers. Metro service providers need to support a higher number of transport protocols and interfaces than long-haul players. They also need to aggregate and switch circuits with a much wider range of speeds (DS-1 to OC-192), and provision and switch circuits between networks much more frequently than in the long-haul segment. Successful equipment vendors will have to acquire or develop emerging component technologies and operating system/network planning software, and cost effectively deliver them to metro service providers through their product offerings. Specifically,

- In the metro transport segment, innovations and technologies that significantly improve service provider efficiency and enable point-and-click flow-through provisioning of service needs to be developed and made available to the marketplace. This should include minimal equipment footprint and costs by combining multiple NEs, such as SONET ADMs, *Digital Access Cross-connect System* (DACS) systems, DWDM transmission systems, and switch/routers. The objective? Improving multiprotocol aggregation/multiplexing and, eventually, increasing flexibility (rapid provisioning) through optical switching, tunable lasers, and tunable filters. These advancements should reduce carriers' capex costs due to savings on the hardware. The advancements should also lower operating expenses via easier provisioning and network management.

- In the metro access segment, GigE is likely to emerge as the primary local access packet transport protocol, given its advantages in three areas: a natural fit into enterprise IT infrastructures, relatively low-cost equipment, and ease of provisioning and upgrading. However, QoS capability is critical in order to expand offerings to include VoIP and IP VPN. These capabilities probably won't be seen in a widespread manner until 2004.

▪ Service providers should deploy SONET-lite/next-generation SONET to incorporate new applications, leverage cost efficiencies, and help reduce the complexity of managing their network.

3.20   The Equipment Vendors

Now that a profile of all the service providers in the metro area Ethernet space has been provided, a profile of some key equipment vendors operating in this space will be presented as well. Not all equipment vendors will be presented here; just those vendors who are having an impact on the marketplace as it relates to (GigE) in the MAN. This listing will be delivered alphabetically and broken down into two sections: legacy equipment providers and upstart equipment providers.

3.20.1   The Upstart Equipment Vendors

Similar to the launch of multiple new service providers to support Ethernet's migration into the MAN and WAN, a number of new equipment startups are also trying to lay their claim in the metro marketplace. It's important to note that just like the service provider segment, it's very possible that some or even all of the vendors listed in the following section will either merge with each other or be purchased by a much larger business entity. When this occurs, many times it is simply so one manufacturer can easily gain access to new technologies that were developed by another manufacturer. It is a means to gain competitive advantage.

3.20.1.1 ADVA Optical Networking   ADVA is a subsidiary of German-based ADVA AG. ADVA targets the metro marketplace with an expansive line of DWDM and CWDM-based optical systems to support metro core transport, metro access, and Fortune 1000 enterprises. ADVA solutions appear to be very scalable, and a wide variety of gear is available to support many different types of metro architectures. For more information, go to www.advaoptical.com.

3.20.1.2 Appian Communications   Appian Communications specializes in optical Ethernet systems and has developed the OSAP. The OSAP optical Ethernet system is designed for both CO collocation and customer premise applications. The system supports 64 Kbps to 1 Gbps data rates and offers SONET interfaces.

Appian is a leader in developing gear that supports the migration of EoS. Appian sees *Ethernet private line* (EPL) as a major cash cow—an application that will be good for VoIP, SANs, video transport, and legacy *Systems Network Architecture* (SNA) traffic. EPL is also considered a high-speed alternative to T1, (nxT1), and DS-3 services—when customers want to make the next leap in capacity at a lower cost than SONET. Appian's solution maps end-user Ethernet streams onto SONET links, maintaining per-flow (per-customer-link) QoS and optimizing SONET for packet-based traffic. This is technically feasible, but security and QoS questions will linger until some real-world proof surfaces that security and QoS are non-issues. However, almost all observers believe there is potential to make this type of application work. Appian also uses a *virtual circuit* (VC) architecture that maps traffic coming in on one port to a VC on another port, so that hackers cannot spoof their identities and access someone else's circuit. In the case of shared traffic on VLANs, Appian plans to create a unique TLAN identifier for each customer VLAN so that membership of a particular VLAN cannot be learned.

3.20.1.3 CIENA Corp CIENA Corp offers two products targeted at the emerging metro market opportunity: MultiWave Metro and MetroDirector. The MultiWave metro is a clear-channel metro DWDM system that supports up to 24 protected channels per fiber and approximately 100 km distances over ring or point-to-point deployments. 10 Gbps line cards are available as well as SONET and data muxing capacity of ($4 \times$ OC-3). Current applications for these products include trunking between COs/POPs, wavelength services, high-capacity data services (GigE), and SAN services.

CIENA acquired MetroDirector through the Cyras acquisition, which closed in March 2001. MetroDirector is a multiservice switch (MSPP) that can support up to 40 Gbps unprotected (16 OC-48 ports) per chassis in one-third of a standard 7-foot bay, and it can groom down to the STS-1 level. The system currently provides optical interfaces including (SONET) OC-3 to OC-192, as well as DS-3 and GigE for data, and traditional voice switching on one platform. The MetroDirector is currently in multiple carrier trials, and CIENA announced contracts for the product with Level 3 in May 2001 and CEC/IDN in China in June 2001. In early 2002, CIENA purchased ONI, another manufacturer of optical-networking products. ONI's key metro optical product is the ONLINE 2500.

3.20.1.4 Extreme Networks Extreme Networks, launched in 1996, is a supplier of next-generation Ethernet switching products based in Silicon Valley. The company has developed a full product line of core, access, and edge network Ethernet switches. The *Black Diamond* product series consists of carrier-class GigE switches for core and access network applications. The *Alpine* series are optical Ethernet switches designed for competitive carrier POPs and data centers. Both product lines offer gigabit wire-speed operations and application-aware switching technology. The *Summit* product line is a series of desktop Ethernet switches designed for a wide variety of edge network applications.

Key: Extreme is becoming best known for the fact that its metro Ethernet equipment can offer "bandwidth by the slice." In other words, bandwidth in 1 Mb increments can be delivered to customers on very short notice. This concept has been used effectively by several of the ELECs who are trying to use Extreme's differentiator to differentiate themselves in the marketplace. And it's working. Who can argue with the disruptive but brutally efficient concept of delivering the equivalent of a T1 of bandwidth in hours (sometimes minutes) when it's needed? Compared to waiting weeks for an ILEC to deliver a T1, this feature is nirvana to enterprises in today's world. How does it work? Competitive carriers (such as ELECs and BLECs) deliver either a 100 Mbps or 1 Gbps Ethernet pipe to the customer's premise. They might initially turn up just part of the pipe (such as 10 Mbps) and sign agreements with the customer to turn up 1 Mb increments of bandwidth as needed for X dollars per month.

3.20.1.5 Foundry Networks Foundry Networks is an industry leader in optical Ethernet switching with its *BigIron* product line of layer 3 backbone switches for MAN and LAN/WAN applications. The systems offer up to 480 Gbps switching capacity and wire-speed IP routing. Foundry also markets the *FastIron* and *EdgeIron* edge network optical Ethernet product lines.

In August 2002, Foundry announced that the company had captured the top market positions for overall layers 4 to 7 web switching, layer 3 10 GigE switching, and layer 3 modular 1 Gbps (1,000 Mbps) Ethernet switching, according to the survey conducted by the Dell'Oro Group, a leading market research firm. The survey also found that Foundry has achieved the number-one position in layer 3 modular 1 GigE switching with 19.3 percent market share of all port shipments, a 100 percent market share of all port shipments in layer 3 10 GigE switching, as well as an unprecedented 79.8 percent market share in the strategic and fast-growing modular chassis-based layers 4 to 7 segment. The 2002 Second Quarter Ethernet Switch Report was released by the Dell'Oro Group on August 14, 2002. This covers all major manufacturers in the Ethernet switch market space. These market share calculations are based on actual port shipments and offer the most accurate measurement of market penetration.

Foundry's Global Ethernet solutions for metro service providers offer an architecture that allows layer 2 networks, layer 3 networks, or MPLS-based networks. For carriers building layer 2 metro Ethernet networks, Foundry offers a comprehensive set of purpose-built features such as 802.1w, SuperSpan, Super VLAN aggregation, and layer 2 over PoS that provide dramatic improvements to layer 2 metro network scalability, availability, and manageability. Foundry's MPLS implementation enables carriers to build and scale metro networks on demand and migrate to MPLS networks seamlessly. With a complete line of layer 2/3 switches, layers 4 to 7 switches, and Internet routers, Foundry provides a single point of purchase, support, and ownership for the entire solution.

SuperSpan uses the concept of virtual hubs and spanning tree domains to scale the layer 2 network. SuperSpan enables a single large spanning tree protocol domain to be organized into a collection of small, easy-to-manage, faster converging spanning tree domains. Each domain generates its own control packets called *Bridge Protocol Data Units* (BPDUs), which are used to decide which of the redundant interfaces should be placed in an active or blocking state for that domain. This is the underlying principle in spanning tree protocol that prevents loops and creates a single forwarding data path. Foundry's BigIron switches join the spanning tree protocol domains together in a loop-free manner while providing high availability. The interfaces on the BigIron switches that connect to each spanning tree protocol domain are defined as boundary interfaces. These boundary interfaces use SuperSpan algorithms to partition the BPDUs in one spanning tree protocol domain from the other. This effectively decouples the spanning tree protocol in one domain from the other, making each domain a self-contained network topology with its own spanning tree protocol.

Foundry has won many industry awards, such as NetWorld + Interop 2000 Atlanta "Best of Show"—Global Ethernet; InternetWeek "Best of Breed"—BigIron 4000; PC Magazine Editors' Choice—FastIron 4802; and Networld + Interop 2002 Tokyo "Best of Show" for the BigIron Switch configured with 10 GigE.

3.20.1.6 LuxN LuxN is a 1998 startup specializing in metro DWDM and CWDM systems. LuxN markets a full line of DWDM products for core and access network applications, including the WideWav DWDM and *Coarse WDM* (CWDM) product lines. LuxN manufactures five key product sets that are capable of delivering metro Ethernet services.

3.20.1.6.1 WavSystem 3200 (WS3200) LuxN's WS3200 series of optical transport solutions enables the rapid deployment of high-bandwidth data, voice, and video services on a simple, intelligent platform.

With a protocol-independent platform, the WS3200 delivers OC-*n*/STM-*n*, GigE, Fast Ethernet, Fibre Channel, or traditional voice traffic over DS-1. The WS3200 fully complements the WS6400 to give service providers the ability to build a cost-effective, end-to-end metro optical transport network.

3.20.1.6.2 The WavSystem 3234 (WS3234) The WS3234 can multiplex up to 16 wavelengths on a single fiber using CWDM technology, and collects and manages optical traffic from the end users' equipment via the WS3202, WS3217, and WS3208. The WS3234 can support a variety of WDM architectures, including hub-and-spoke, point-to-point, multidrop, and ring configurations, at distances up to 250 km. The WS3234 complies with NEBS/*European Telecommunications Standards Institute* (ETSI) standards, allowing for collocation in a carrier environment.

3.20.1.6.3 The WavSystem 3217 (WS3217) The WS3217 can multiplex up to eight wavelengths on a single fiber using CWDM technology and is designed for use at either the service provider's POP or a CPE location. Up to eight remote WS3217s, WS3208s, or WS3202s can be connected to a POP-based WS3217 in a hub-and-spoke, point-to-point, multidrop, or ring configuration. The WS3217 also complies with NEBS/ETSI standards, allowing for collocation in a carrier environment.

3.20.1.6.4 The WavSystem 3208 (WS3208) The WS3208 supports up to four wavelengths in only two *rack units* (RU) of chassis height. The WS3208 provides an intelligent and flexible connection between enterprise applications, including SANs and high-speed network services.

3.20.1.6.5 The WavSystem 3202 (WS3202) The WS3202 supports a single wavelength and is the CPE component of LuxN's WavSystem. Occupying one RU, the WS3202 provides a demarcation point between the enterprise and carrier portions of the network.

Figure 3-10 shows the WavSystem family of metro Ethernet products. The graph depicts the appropriate place for the systems in the network.

3.20.1.6.6 Topologies and Protection Schemes The WS3200 platform supports multiple topologies such as hub-and-spoke, point-to-point, multidrop, and ring configurations. In addition, multiple protection schemes are avail-

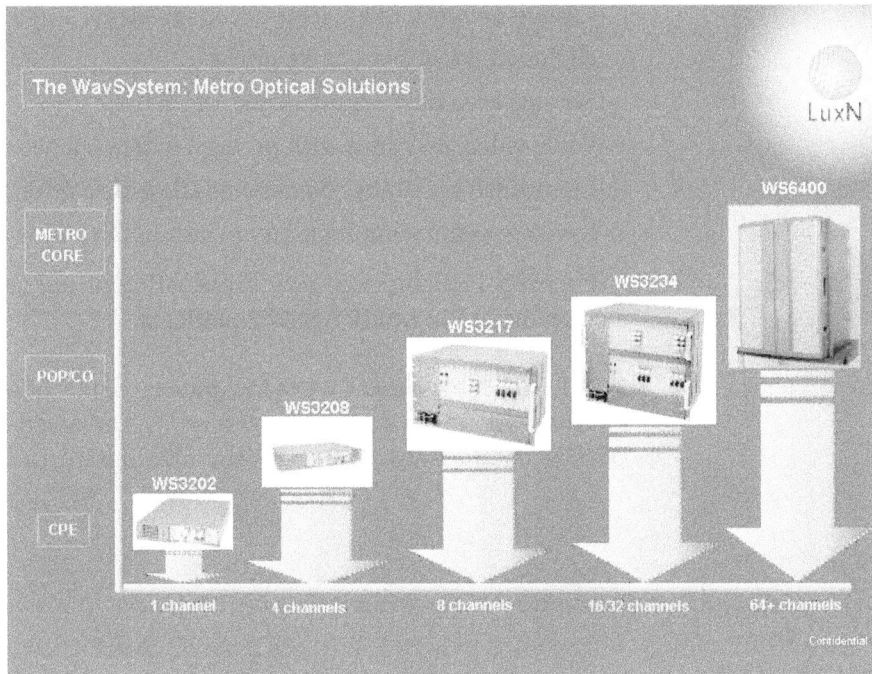

able such as optical *Unidirectional Path Switched Ring* (UPSR) (path only or path + hardware), optical UPSR (single-fiber bidirectional protection), optical mesh topology protection and restoration, layer 2/3 and SONET/SDH protection, and optical path protection.

3.20.1.6.7 Applications LuxN's WS3200 can be used in the following application scenarios:

- Native GigE connectivity services
- TLAN services 10/100 BaseT Ethernet
- Transport of high-speed storage services
- Mixed video and data transport
- High-speed Internet access
- DS-L and wireless traffic back haul
- Optical aggregation and switching
- Fiber exhaust
- SLA and performance monitoring at the optical level

- Unique network management capabilities
- High-speed VLANs—at T1 prices
- Storage area networking across the WAN
- Voice, video, and data with protocol independence
- Bandwidth scalability on demand (flow-through provisioning)
- Revenue-generating high-bandwidth applications and services

Figure 3-11 illustrates how these different applications can be implemented using the LuxN WS3200 platform.

3.20.1.6.8 DWDM and CWDM Integration LuxN offers the industry's first solution with an option for aggregating CWDM and dense ITU wavelengths DWDM on the same fiber, thereby eliminating the traditional trade-off

Figure 3-11
An implementation of LuxN WS3200 family of products in metro environment

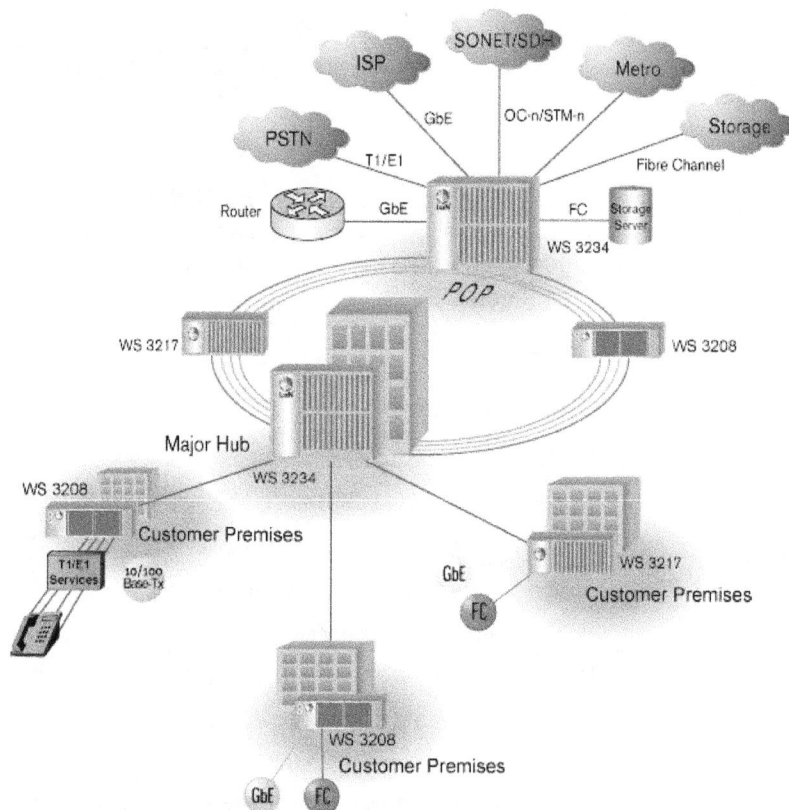

between low cost and wavelength scalability. When used together, LuxN's solution supports 20 wavelengths over a single fiber pair: 4 coarse wavelengths and 16 ITU wavelengths.

3.20.1.6.9 *ColorSIM and Service Level Agreements (SLAs)* ColorSIM is a patented LuxN software system that's a per-wavelength, protocol-independent, nonintrusive monitoring technique that enables a carrier to offer SLAs for any protocol. ColorSIM provides a truly unique performance monitoring technology that reports optical signal-link integrity and QoS metrics while continuously running live traffic. ColorSIM enables service providers to measure the true *bit error rate* (BER) by utilizing an analog-signal monitoring technique that operates directly on the physical symbols that represent bits.

3.20.1.6.10 *ColorValve* ColorValve enables service providers to remotely adjust bandwidth on a per-wavelength basis. Through the use of a web browser, either the carrier or end user can upgrade the bandwidth on a temporary or permanent basis. This feature can result in considerable cost reduction through the elimination of truck rolls, technician time, and physical site upgrades. LuxN has major contracts with AT&T, TWT, and several other major carriers.

3.20.1.7 Riverstone Networks Riverstone Networks, another gutsy and impressive startup, specializes in next-generation optical Ethernet systems. The RS Metro Router product family supports GigE over DWDM, broadband cable, ATM, and PoS while also supporting prestandard RPR and MPLS.

The RS32000 intelligent edge router aggregates and routes network traffic at service provider facilities and collocation centers before it is routed to the Internet backbone. The RS38000 product family of optical metro aggregation routers support a full range of Internet routing protocols and is capable of aggregating all IP access technologies including 10 GigE, GigE, Fast Ethernet, WDM, PoS, ATM, and channelized DS-1. The RS16000 offers high GigE port density and supports 10 GigE. The product supports 12 ports per rack unit, or 540 ports per 7-foot rack.

The RS8000 and RS8600 are small form factor metro service routers that are used as Internet access or network traffic aggregation platforms. The RS3000 extends service delivery to the metro access edge, delivering services over any infrastructure: WDM, GigE, DS-1, DS-3, SONET, and ATM. The RS2000 and RS2100 are flexible metro access routers that support traditional connectivity along with optical modules.

Riverstone Networks is focused on the metro opportunity, and has sold products to Qwest, British Telecom, Hutchinson, AT&T, Japan Telecom, Strom, Videotron, Completel, Retevision, Shanghai Post, UPC, Cox, Telia, UUNet, Dacom, Qwest, NTT (Japan), and Telefonica (Spain).

In July 2002, Riverstone Networks announced that Dreamline Corporation, one of Korea's leading broadband data service providers, has deployed Riverstone's RS routers in an end-to-end metro Ethernet solution to provide broadband Internet access, virtual private LAN services, and online multimedia and hosting services to its 1 million business and residential subscribers.

Key: According to a recent report by industry analyst firm RHK, Riverstone is the largest provider of metro solutions with a 41 percent share of the worldwide market.

Riverstone is also in trials with BellSouth. They've signed major agreements with many municipalities and phone companies around the world.

This aggressive company has attempted to break into the incumbent market through an agreement with incumbent darling Tellabs. It's believed that other Ethernet equipment vendors will follow this model of partnering with an established equipment supplier, or, depending on their long-term financial viability, will look to be acquired by these mature players. This gutsy, savvy startup is worth watching as they give Cisco a run for their money in the marketplace.

3.20.2 The Legacy Equipment Vendors

The legacy equipment vendors are those vendors who have been around for a long time (more than 20 years) in some form. These companies have been known for their huge bases of employees and their huge bases of customers. They have retained most of their customer base over the years, but since the financial collapse in the telecom industry that began in 2000, most of these companies have shed tens of thousands of employees because they have massive idle inventories and a dearth of orders from their largest customers. The swagger these companies once had has largely disappeared as they've been humbled by the telecom meltdown and the lack of willingness to buy on the part of their largest customer base, the RBOCs. Nortel and Lucent stocks were both trading at less than one U.S. dollar on October 14, 2002. Nortel's employee base has shrunk from around 90,000 down to

35,000 as of October 2002. Similarly, Lucent has been laying off employees in chunks of 10,000 since 2000—so has Motorola. Lucent's most recent layoff of 10,000 employees occurred on October 9, 2002.

Nevertheless, the big boys in the equipment space are still trying to keep themselves inserted in the metro marketplace because they understand that it's still one of the hottest markets around and will continue to be for the next few years at least. They don't want to be left behind, especially at a time when their revenues—overall—are so anemic.

3.20.2.1 Alcatel The Alcatel 1650 SMC and 1660 SM are next-generation SONET multiservice nodes for metro applications. Alcatel also markets DWDM systems for metro network applications. The Alcatel 1686 is a 32-channel DWDM system and the Alcatel 1696 metro span DWDM system is designed for regional networks. Alcatel has an *original equipment manufacturer* (OEM) contract with ADVA for metro DWDM systems. (This means that Alcatel makes the gear and ADVA slaps its name on the gear.) The company also markets a full line of optical termination products, gateways, and optical cross-connects. All these devices could supplement metro Ethernet transport.

3.20.2.2 Cisco Systems Cisco has gained increasing traction in the metro market, particularly with some incumbent carriers. Cisco has three main metro products: the 15327, 15454, and 15540. The 15327 provides SONET/SDH transport, optical networking and multiservice on demand. This product aggregates data, voice, and video service. It supports services such as TDM, Ethernet, and ATM, and provides integrated data switching and cross-connect functionality. Through its acquisition of Cerent, Perelli, and Qeyton Systems, Cisco markets a line of next-generation SONET systems for metro applications.

The ONS 15454, originally a Cerent product, is a leading metro ADM that appears to be gaining increased traction with service providers. The systems support TDM, IP, and ATM services over 18 wavelengths. The ONS 15454 is a multiservice system, combining SONET ADM, cross-connect, data, and optical switching onto a single platform. It can operate up to 15 Gbps of unprotected bandwidth in one-fourth bay. Available interfaces include OC-3 to OC-192, DS-3, DS-1, 10/100 Mbps, and 600 Mbps Ethernet. The 15454 also contains a VT1.5 and STS-1 grooming card. Announced customers include a large number of incumbent and emerging carriers. The ONS 15540 is a metro DWDM system that can manage up to 32 protected wavelengths per fiber. This box also has OC-3 to OC-48 interfaces.

3.20.2.3 Lucent Technologies Lucent introduced its suite of metro transport and switching products at SuperComm in 2001. The Metropolis portfolio consists of the Metropolis metroDMX, the metroMSC, the metroEON, and the metroMIS. The Metropolis suite is deployed or in trials with 18 customers.

The metroDMX is a next-generation SONET ADM with both optical and electrical interfaces for service aggregation at the network edge. Additional services include 10/100 Mb and GigE line cards, and the system supports a total capacity up to 40 Gbps in one-fourth of a 7-foot bay (with VT1.5 grooming).

Key: It is believed that this product gives Lucent an offering similar to the Cisco 15454 platform, because it's a SONET ADM platform with integrated data interfaces.

The metroMSC evolved form the Chromatis acquisition, which was announced in May 2000. The metroMSC offers SONET and data aggregation, a hybrid ATM/TDM switch matrix, a TDM-only switch matrix, and integrated DWDM for trunking purposes. The hybrid switching matrix supports up to 70 Gbps capacity for the TDM fabric and 23 Gbps for ATM switching, while the TDM-only matrix provides up to 70 Gbps and grooms down to the STS-1 level. The chassis occupies one-half of a 7-foot bay and has 28 universal service slots that can each be populated with optical elements (ADM modules or optical channel modules for OEO conversion) or tributary interfaces (line interfaces such as GigE, DS-3, OC-48, and so on). The system's ATM switch matrix enables the metroMSC to efficiently aggregate ATM traffic such as private-line data and *Digital Subscriber Line Access Multiplexer* (DSLAM) traffic, much like a traditional ATM switch.

The Metropolis metroEON is a newly created clear-channel metro DWDM system that is an adaptation of Lukens' long-haul DWDM offering —the OLS 40G systems. Their systems support 32 protected channels with 100 percent add/drop capabilities and OC-3 to OC-48 optical interfaces. The metroEON also supports GigE and other non-SONET line cards for distances up to ~384 miles in ring or point-to-point deployments.

This system can also be deployed with the Metropolis metroMLS. The metroMLS is a smaller system used for metro access applications to deliver greater-distance wavelengths to the network edge through CWDM. CWDM is a technology that spaces optical wavelengths wider than DWDM, using cheaper components for smaller COs or even large enterprises/MTUs in some applications. In that application scenario, it would be similar to ONI's ONLINE 2500 or Redback's SmartEdge 100 platform.

3.20.2.4 Nortel Networks Nortel is a dominant supplier of optical-networking equipment worldwide. The company markets the OPTera product line for metro networks. In 2001, Nortel separated its metro optical division from the original optical division, which now only consists of Nortel's long-haul optical products. Nortel's metro optical flagship products include the OPTera Metro 3500, OPTera 5100, and OPTera Metro 5200. The OPTera Metro 3500 is Nortel's SONET ADM offering and provides interfaces supporting a wide range of services including TDM (DS-1 to OC-48), ATM and GigE. The OPTera Metro 5200 is Nortel's metro DWDM system that offers 32 protected or 64 unprotected wavelengths. It can be used in both a point-to-point and a ring topology. Nortel states that the open and modular architecture of the OPTera Metro 5200 delivers network scalability, per-wavelength manageability, protocol and bit-rate independence, and ring survivability. The 5200 is actually an MSPP as it can transport the following protocols in their native formats end to end: Fibre Channel, ESCON, FICON, GigE, SONET, 10 GigE, and D1 Video.

Nortel's Passport 8600 multiservice switch is another element of Nortel's metro optical Ethernet strategy. Service providers can use the Passport 8600 to deliver GigE services in three network architectures: Ethernet over DWDM, EoS, and Ethernet over RPR. The Passport 8600 offers QoS-based traffic prioritization, scalability, and a range of interfaces including Fast Ethernet, GigE, PoS, and ATM.

In 2002, Nortel unveiled its largest multiservice switch yet—the long-awaited Passport 20000, which supports ATM and MPLS. The platform is in trials and will offer 160 Gbps of switching capacity and a variety of 10 Gbps interfaces, including OC-192 PoS, a 10 GigE interface, and four OC-48 ATM links. The previous high-end Nortel switch, the Passport 15000, topped out at a capacity of 40 Gbps. Unlike the Passport 15000, the 20000 is also designed specifically for the core of carrier's networks. The Passport 20000 will let carriers link a variety of services—including traditional voice, wireless, IP data, IP VPNs, ATM, and frame relay—into a single backbone network.

Carriers can use the 20000 to migrate directly to an IP core, or continue using ATM in the backbone. With either technology, MPLS can be used to provide layer 2 IP VPNs over an ATM core and traffic engineering over an IP core. For carriers using Nortel DWDM in the core of the network, the Passport can dynamically set up and tear down individual wavelengths as capacity is required, according to Nortel. A separate Nortel transport device on the edge of the optical infrastructure will then convert the electronic signals to light waves for transmission.

Nortel has not been immune to the pressures of the marketplace. Nortel has been struggling since early 2001, and the company's struggle shows no signs of abating as of early 2002. The company shed tens of thousands of employees from mid-2001 to early 2002. On July 25, 2002, Nortel stock closed at 82 cents.

3.20.2.5 Tellabs Tellabs has intensified its focus on the metro market through its TITAN 6100 metro DWDM system. This DWDM platform currently offers OC-3 to OC-48 interfaces and supports up to 32 protected wavelengths over distances up to 100 km (~60 miles) for ring, mesh, and point-to-point deployments without manual or automatic power tuning needed by most systems to rebalance signals when wavelengths are added. Non-SONET interfaces and SONET multiplexer cards are expected to be available in 2002, including GigE and ESCON cards and subrate OC-3/OC-12 multiplexers. Tellabs management announced in mid-2001 that initial revenues for the product have been recorded for the June quarter and the systems are currently in trials and running live traffic with multiple service providers. It is expected that Tellabs will package this platform with its family of cross-connect systems (TITAN 6700, TITAN 6500, TITAN 5500, and TITAN 5321) in order to provide carriers with optical transport and efficient grooming for intracity and metro core networks. Table 3-7 shows a breakdown of the metro equipment players and their key customers.

3.20.3 Summary: The Equipment Space

Legacy SONET/SDH networks will be expanded and upgraded at least into 2004 and 2005. Newly constructed networks based on next-generation optical network systems offer the best hope for future profitability. These networks will be designed specifically for data-centric traffic. They'll lower operating costs, improve packet-handling efficiency, and position network operators to capture revenue from new services.

For equipment vendors, the metro Ethernet marketplace may become a battle of competing products within specific customer segments more than a battle between competing technologies. Customer share will be more important than market share.

On-time delivery of working product and subsequent version enhancements in key product niches will ensure market success.

Key: Vendors must be very careful not to mislead savvy customers—the service providers—who thoroughly test vendor equipment prior to making major purchasing decisions and deployments into their networks. Some

Table 3-7 Equipment vendor-to-service-provider cross reference

Metro DWDM	U.S. Customers
Alcatel 1696 Metro Span	None disclosed
Centrepoint Broadband Technologies Inc. (Zaffire) Z3000	FiberStreet
Cisco 15200	
Cisco ONS series	Cambrian Communications and Cogent Communications
Lucent Metropolis MLS	None disclosed
LuxN WaveSystem	TWT and Yipes Communications
Marconi Communications SmartPhontiX PMM	None disclosed
Network Photonics CrossWave	None disclosed
Nortel OpTera Metro	MFN, SBC Communications
ONI Systems Inc. ONLINE	Sphera Optical Networks, AT&T
Siemens Optisphere Networks TransXpress Waveline	None disclosed
Sorrento Networks GigaMux	TeraBeam and GigX Communications
Tellabs Focus 6100	None disclosed

Sonet Optical Transport Platform	U.S. Customers
Alcatel 1603 SMX	AT&T, WorldCom, and Sprint
Cisco ONS 15454	Cambrian Communications, Cogent Communications, Looking Glass Networks, and Sigma Networks
Lucent Technologies WaveStar	XO Communications
Nortel Networks S/DMS TransportNode	CenturyTel and Prism Communications Services
Tellabs Titan 5500	Qwest Communications

Table 3-7 Equipment vendor-to-service-provider cross reference (continued)

Multiprotocol Transport Platform/ Next-Generation SONET	U.S. Customers
ADC Telecommunications Cellworx STN	Iowa Communications Network and Mid-Continent Communications
Atoga Systems OAR	None disclosed
Dynarc 5116	None disclosed
Fujitsu Flash 4500 (formerly Flash 2400ADX)	None disclosed
Lucent Technologies Metropolis MSX series	Cogent Communications
Luminous Networks PacketWave Series	Qwest Communications
Metro-Optix CityStream line	1sTel, Qwest, and Network USA
Native Networks EtherMux-M	None disclosed
Nortel Networks OpTera Metro 5200	Giant Loop
Redback Networks SmartEdge series	Ace Communications Group and Qwest
Tellabs Titan 6500	Sprint
White Rock Networks VLX Series	None disclosed

Metro Optical Ethernet switches and aggregators	U.S. Customers
Cisco Catalyst series	Cogent Communications and MFN
Extreme Networks Summit, Black Diamond, and Alpine	IntelliSpace, Cogent Communications, and Yipes Communications
Foundry Networks BigIron series	Stream Intelligent Networks, Telseon
Riverstone Networks RS3000, RS3800	IntelliSpace, Telseon

Source: Steven Titch. "GigE Evolution," *America's Network Magazine*," January 15, 2002.

vendors' equipment simply doesn't function as advertised, or the vendors will string customers along for months or years promising that currently advertised features of certain equipment will "actually be available with future software releases." The problem is that many times, these future software releases themselves either don't perform as advertised, or they're constantly delayed. Nothing will damage a vendor's credibility faster than putting a customer through this type of pain.

Expect additional mergers, acquisitions, and consolidation of equipment suppliers, especially if deployment of broadband services continues its slow growth.

After reviewing the product suites of all the equipment vendors listed in this section, a trend becomes self-evident. To accommodate legacy infrastructures while simultaneously adapting their product lines for the future —which many say will be Ethernet-based—all the vendors reviewed are incorporating both SONET and Ethernet capabilities in their current product lines.

Is GigE a business model or just a service? After examining the information supplied thus far, the answer to this question is simply yes.

Competing Technologies

4.1 The Market Impact of Gigabit Ethernet (GigE)

The presumed advantage of Ethernet is that it can essentially stretch the cheap and reliable local computer network out to meet the Internet's fiber-optic backbone. Where this is possible, it helps eliminate the need for intermediate layers of gear using the *Asynchronous Transfer Mode* (ATM), frame relay, and *Synchronous Optical Network* (SONET) protocols. The rigidity of SONET and ATM versus the flexibility of all forms of Ethernet is a major selling point for Ethernet's placement in carrier networks. But SONET- and ATM-based services are up and serving customers, and are not likely to go away anytime soon. Even the finest technology solutions must answer to the corporate "spending gods." During the telecom-economic implosion of 2001 and 2002, hardly any service or equipment business was immune to the cost cutting and constriction of capital spending.

As we saw in Chapter 3, "The Metro GigE Marketplace," businesses in the top 20 major metropolitan areas can currently purchase 100 Mbps of Internet connectivity for $1,000 a month from emerging carriers such as Cogent, if they are fortunate enough to be located near Cogent's fiber-optic cabling infrastructure. Customers do not have to purchase expensive ATM or SONET interfaces to accommodate higher bandwidth requirements, and the service is compatible with a standard Ethernet port on existing routers and/or switches. Network managers do not have to train their operations groups in arcane one-off technologies because Ethernet is ubiquitous and relatively simple. Any networking architecture that runs Ethernet over another network technology adds complexity and overhead. It creates yet another layer to be managed in the network. Each network layer requires separate equipment, separate operational and management frameworks, and probably a separate support staff. Finally, organizations can opt to purchase only the bandwidth used from service providers that offer usage-based *Gigabit Ethernet* (GigE) services, instead of being forced to waste fixed slices of voice bandwidth for bursty data traffic.

Key: By reducing the incidence of other protocols in the network, transport architectures are flattened and simplified, leading to lower costs with regard to service delivery. The rigidity of SONET and ATM versus the flexibility of all forms of Ethernet is a major selling point across a number of the GigE providers.

Because metro optical Ethernet is primarily a point-to-point technology as of 2002 (only one or two *Ethernet Local Exchange Carriers* [ELECs] offer multipoint Ethernet), IDC expects it to have minimal effect on the frame relay and ATM markets in the near term. However, Ethernet's impact will still be felt in the marketplace—count on it.

The traditional data transport services that are covered in this chapter aren't going away anytime soon. As a matter of fact, their *compound annual growth rates* (CAGRs) are still healthy, and they'll continue to grow into 2006. The important thing to remember is that these growth rates may decrease between 2002 and 2006 if metro (and *wide area network* [WAN]) Ethernet services take hold as expected. Table 4-1 shows the estimated CAGR for all the services profiled in this chapter.

4.2   Ethernet Versus SONET

In terms of features and functionality, SONET is the one technology that presents Ethernet with the biggest challenge in terms of market displacement and "cannibalization" of SONET sales and revenue.

4.2.1   The Constraints of Legacy SONET

The debate about SONET versus Ethernet in the MAN is as strong as ever, as upstart MAN equipment vendors and service providers praise the

Table 4-1

Estimated growth rates for legacy data transport technologies and GigE

Service	Estimated CAGR 2001 to 2006
DS-1	2.0%
DS-3	8.0%
OC-3	38.0%
Local frame relay	12.4%
Long-distance frame relay	14.3%
ATM	26.6%
GigE	26.0–37.0%

virtues of Ethernet while berating the shortfalls of SONET. With data traffic and demand for data services growing, it has become clear to many in the industry that the existing, multibillion-dollar *Incumbent Local Exchange Carrier/Competitive Local Exchange Carrier* (ILEC/CLEC) legacy SONET infrastructure is incapable of handling this load. A quick overhaul of this SONET landscape is unlikely, but evolving technologies such as optical Ethernet will chip away at SONET's dominance.

Within today's voice-oriented networks that are based on *time division multiplexing* (TDM), SONET technology has become widely deployed to provide high-capacity transport with the capability to scale up to gigabit per second rates. OC-12 rates are equivalent to 622 Mbps. OC-48 rates are equivalent to 2.48 Gbps. OC-192 rates are equivalent to 9.953 Mbps (almost 10 Gb). SONET/*Synchronous Digital Hierarchy* (SONET/SDH) self-healing rings enable service-level recovery within tens of milliseconds in the event of a network failure. All these features are supported by well-established standards that enable a high degree of multivendor interoperability. However, with the paradigm shift from voice-optimized, circuit-switched networks to data-optimized, packet-switched networks, the SONET legacy infrastructure is now becoming constrained by its inherent limitations.

Key: Voice traffic is increasing at a rate of 8 to 10 percent a year, but packet data traffic is increasing at a rate of 200 percent a year!

Figure 4-1 shows the huge disparity between the growth rates of voice services versus data services as of 2002.

SONET is basically a multiplexing technology, which means that it enforces the rigid TDM telecom hierarchy. This works well for all-voice traffic as latencies and bandwidth are easily guaranteed by assigning a specific time slot of bandwidth to each connection. However, SONET lacks the flexibility and smooth scalability that's required for data traffic that is inherently bursty. The addition of another layer of overhead encapsulation that is necessary for *Internet Protocol* (IP) traffic also consumes more of the network's available capacity.

Key: In order to conform to the SONET infrastructure when transporting data, carriers have to cram data streams into arbitrarily rigid channels. The slowest channel, VT-1.5, runs at 1.7 Mbps, whereas the next higher increment, STS-1, jumps to 51.84 Mbps. Inherent inefficiencies are created because any unused capacity in each channel is wasted. To make

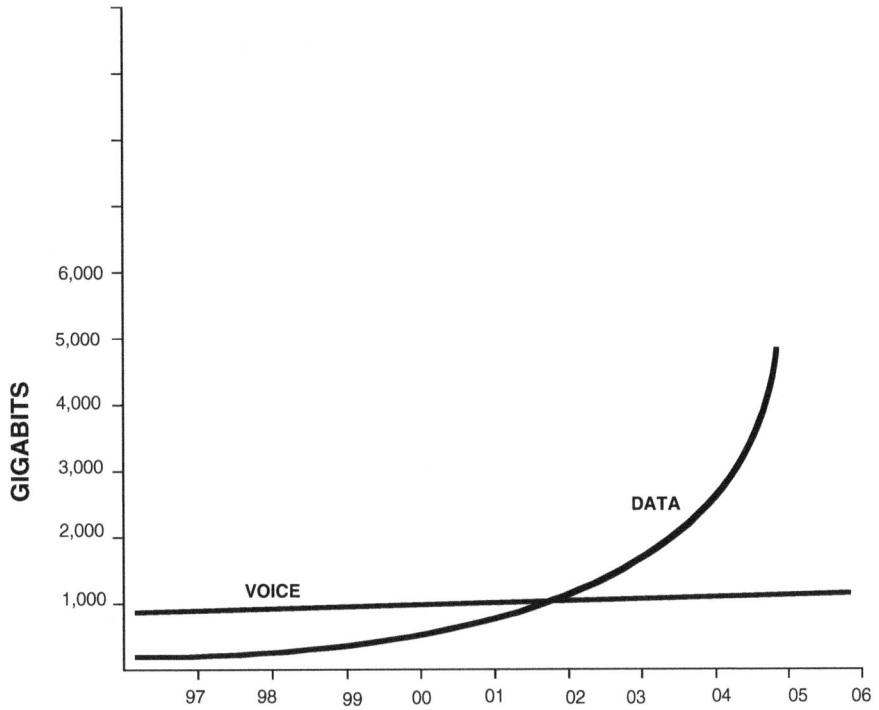

matters worse, these inefficiencies become more pronounced as the ratio
of data traffic to voice traffic has dramatically increased—most of the traffic growth in today's networks is due to data applications. Studies performed in 2002 indicate that data traffic outnumbers traditional
TDM-based voice traffic by a factor of 10:1 in the *Public Switched Telephone Network* (PSTN). However, metropolitan area connectivity has traditionally been provided by SONET rings, which were designed and built
to carry voice traffic.

Despite numerous claims that SONET was doomed and would gradually
be phased out as *dense wave division multiplexing* (DWDM) entered the
market, SONET is still growing. The reason is fairly simple: Direct connection of customer network systems or data networking equipment to DWDM
systems has yet to take place on any significant scale. The result for now is
that SONET networks expand as data networks expand. SONET continues
to be the dominant transport mechanism between DWDM transport systems and switching, routing, and aggregation equipment.

Key: In some carrier networks, core IP routers are being directly connected to DWDM systems through *packet over SONET* (PoS) interfaces, but this has led to little reduction in demand for SONET multiplexers. However, as the cost of WDM components declines, the cost of WDM multiplexers will also go down. When this activity reaches the point where PoS over WDM is more cost effective than SONET, and *quality of service* (QoS) issues related to *voice over IP* (VoIP) are effectively addressed, the demand for legacy SONET equipment will begin its decline.

4.2.2 The Lack of Smooth Scalability and the High Cost of Expansion

Because of the high cost of SONET equipment, service providers face difficult challenges when it comes to scaling their *metropolitan area networks* (MANs) to accommodate the ongoing upsurge in data traffic. The prospect of throwing more SONET cross-connects and *add/drop multiplexers* (ADMs) at rising data traffic levels becomes prohibitively expensive. The requirement to terminate all individual WDM wavelengths with high-priced SONET equipment can also be expensive. By contrast, new-generation, high-speed Ethernet switches offer a major cost reduction compared to SONET ADMs.

In SONET networks, bandwidth is allocated in relatively large, rigid increments—1.5 Mbps, 51 Mbps (OC-1), 155 Mbps (OC-3), and so on—that force customers to wait until their budget can cost justify the next service upgrade when they need more network capacity. Once an order is placed for SONET service, it typically involves multiple service installation truck rolls to provision new circuits. As a result, SONET solutions suffer from long provisioning lead times (several months) and high deployment costs.

The fixed TDM channel allocated to each service in a SONET system is wasted on bursty packet data flows. For example, if a subscriber requires more than two DS-1 circuits for Internet access, carriers typically install a DS-3 access circuit (45 Mbps) that uses a full STS-1 payload across the SONET network.

Figure 4-2 shows the complexity involved in SONET systems, from the protocol overhead to the number of expensive SONET ADMs involved in the network topology.

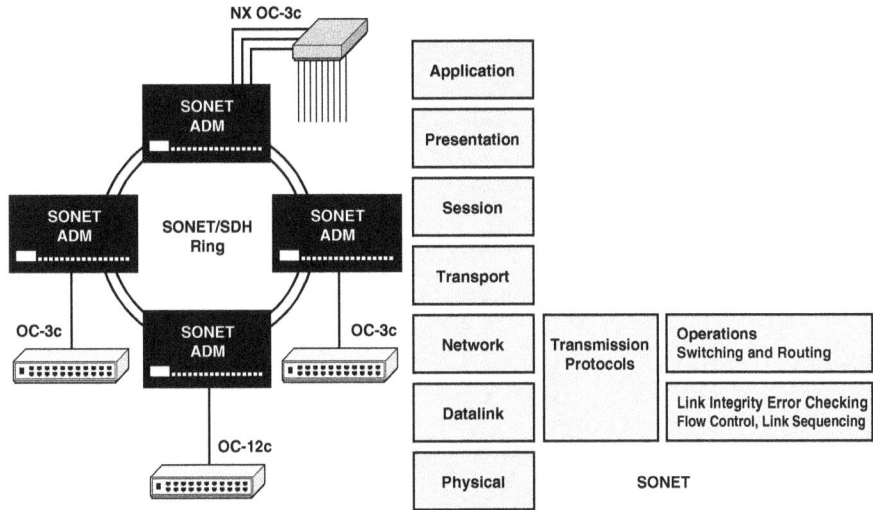

Figure 4-2
Typical SONET
infrastructure

Key: Optimal bandwidth utilization for voice and data services cannot be guaranteed over SONET since unused bandwidth in a given STS-n pipe is wasted. By bolting a router or an Ethernet switch on top of a SONET ADM, the cost and complexity associated with these multiple technologies is not eliminated.

4.2.3   Restoration Versus Efficiency

Four-fiber *Bidirectional Line-Switched Rings* (BLSRs) in the SONET world are implemented with working rings (one ring for transmit, one ring for receive) and separate protection rings (one ring for transmit, one ring for receive). One entire pair of rings is sitting completely idle, waiting for a major malfunction (a fiber cut or equipment failure) to disable the other pair so that traffic can be rerouted to the other pair, which occurs in 50 ms or less. Up to 50 percent of network capacity is used for protection in SONET ring-based systems. This failsafe backup strategy doubles the amount of fiber required to operate a SONET ring—fiber that sits idle waiting for an emergency to occur. Figure 4-3 shows the topology and redundancy function of a 4-fiber BLSR SONET ring.

Figure 4-3
Four-fiber BLSR
SONET ring
protection

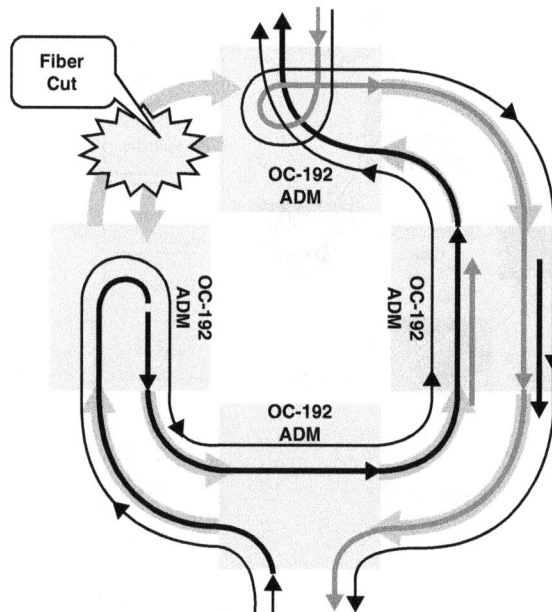

Key: This is one of the main disadvantages of the SONET standard. Fully redundant rings are costly and do not make efficient use of network capacity. On the upside, however, it also offers a simple recovery mechanism that guarantees premium, uninterrupted service when failures also occur.

In SONET *Unidirectional Path-Switched Rings* (UPSRs), traffic flow is unidirectional. A channel provisioned between two points is dedicated to the exclusive use of a particular customer. There's no way to share bandwidth across multiple customers, so in this scenario the cost per megabit per second is extremely high for packet services. Neither SONET approach —BLSR or UPSR—is optimized for data traffic and new emerging Ethernet services.

The ILECs are challenged with figuring out how to replace existing voice revenues with innovative data services and how to simultaneously leverage their legacy networks that reflect a sunken investment of several trillion dollars.

4.2.4   Long Provisioning Cycles and Inflexible Billing Mechanisms

One of the key limitations that puts SONET-based service providers at a competitive disadvantage are the inherently long cycle times and lack of flexibility involved with provisioning SONET connections for customers.

Because SONET was originally designed to efficiently carry uniform voice traffic, it doesn't include the flexible billing and management features that carriers need to offer value-added data services such as *virtual private networks* (VPNs), web hosting, or collocation. Simply providing connectivity does very little for the service provider's bottom line. The ability to offer tiered services maximizes profit.

Key: Without value-added features, service providers who are tied to SONET legacy infrastructures have no built-in mechanisms to quickly recover their investments in network upgrades to support the explosion of bandwidth requirements.

4.2.5   The Burdensome Complexity of SONET Muxing

SONET standards were developed by the *Consultative Committee for International Telegraph and Telephone* (CCITT) (now the *International Telecommunications Union* [ITU]) in the 1980s to overcome the timing complexities, proprietary management, and proprietary optical interfaces associated with the plesiochronous multiplexers and optical terminal products of the time. However, SONET only offered better technical solutions for existing voice-dominated services. It did not offer significantly better access for new services such as high-speed packet data, even though high-speed data was in the early stages of development. SONET's hierarchical multiplex structure combined with the need for *virtual containers* to transport high-capacity services has proven to be a very complex and costly solution —especially for MAN applications where ADM access is required. The complexity of SONET networks is also augmented by the amount of equipment required for end customers to access and use them.

4.2.6 What Are the Options?

Despite having significant advantages for voice traffic, SONET is ill suited to handle the increasing volume of multiservice, data-centric traffic in the metro network. Yet network service providers have a huge investment in SONET equipment that is not fully depreciated. Clearly, SONET will remain a dominant technology in metro networks for many years to come. The question is how will metro networks evolve to meet market demands that now focus primarily on data transport?

Today's challenge for carriers (and Fortune 100 companies) is dealing with traditional SONET/SDH infrastructures that do not support the granularity, scale, low cost, and on-demand bandwidth required by bursty packet applications. As a result, carriers have seen lower profitability as the operating costs and capital investments associated with data services continue to rise.

The options are straightforward: Build an entirely new data services overlay network, leverage the existing infrastructure, or pursue both options at once, albeit gradually. Incumbents will not conduct wholesale replacement of existing SONET systems with optical Ethernet technology any time soon. To optimize investments already made in transport and back-office systems, many carriers are choosing to bring the economics of Ethernet together with their existing SONET network to create a new generation of carrier-class Ethernet services that offer

- Granular, bandwidth guarantees managed by a *service level agreement* (SLA)
- Rapid, even on-demand service activation
- SONET resilience and manageability
- Services that span both the metro and wide areas
- High-speed migration for current data services
- A simple strategy to sell more new services
- Integration with existing TDM services (such as SONET)
- Greatly reduced operating and capital costs

Key: Between 2007 and 2010, unless next-generation SONET and/or *Resilient Packet Ring* (RPR) systems are widely adopted (see Chapter 5, "Complementary Protocols"), the migration away from SONET-based optical services will begin as the legacy SONET infrastructure becomes fully depreciated. This is even more probable (and may happen quicker) if standards under development (such as RPR) are adopted widely in the carrier space.

Following lengthy product testing, ILECs will deploy next-generation SONET or optical Ethernet systems in access network applications. These multiservice systems will be used in conjunction with Ethernet edge networking products.

4.2.7   Ethernet as a SONET Replacement?

There's a lot of talk about GigE these days, especially in the metro, but increasingly in the WAN as well. What should you make of this? Is SONET really going away? Not likely, at least until the legacy SONET plant fully depreciates, and that's years away. Interdependence is the more likely outcome. That said, GigE does make a lot of sense in the metro, where it has taken off at a very difficult time for the telecom market. Carriers such as Cogent Communications, Yipes Enterprises, and Telseon have delivered proof of concept that GigE makes a lot of sense as a metro transport medium and GigE's advantages have become quite clear. Ethernet is becoming a generally accepted protocol in MANs (and WANs) as of 2002. 2002 has been predicted to be the inflection point year in which many early adopter MIS and network managers will choose *Ethernet over SONET* (EoS) in their enterprise networks.

Ethernet technology hasn't quite matured to SONET's level yet, but that's part of the RPR initiative. Once Ethernet is fortified with RPR, the debate between Ethernet versus SONET will probably intensify. The next debate will probably move to ring architectures versus mesh architectures, and the entire process will repeat itself. "The world is moving away from rings and toward mesh topologies," according to Alan Brind, Aura's senior vice president of worldwide marketing.

In fact, two major standards initiatives are under way that should provide a major leap forward in resiliency and QoS for metro area GigE. RPR aims to give Ethernet SONET-level protection and reliability with a data link layer optimized for packet traffic in the *local area network* (LAN), MAN, and WAN, and standards developments such as *Generic Framing Procedure* (GFP) (covered in more detail in Chapter 5), which maps Ethernet packets to SONET transmission links. Both of these standards efforts should be watched closely. (See Chapter 5 for additional information.)

A great deal of interest surrounds the deployment of Ethernet in MANs as a replacement for SONET. Simultaneously, Ethernet advocates are trying to elevate Ethernet in the consumer arena to become a broadband alternative to xDSL and cable access services.

4.2.8 The Disadvantages of GigE over SONET

Metro networks were originally designed for voice transport and had copper last mile muxed onto the core of the metro network. Therefore, the strength built into such networks was the ability to provide dedicated, low-latency connections that had high levels of survivability due to SONET's inherent protection switching feature.

However, since the mid-1990s, it has become clear that SONET has major disadvantages as the primary layer 1 carrier protocol:

- Ethernet transport interworks very well with IP traffic, and IP is the fastest-growing traffic form in metro networks. As applications continue to migrate to IP, Ethernet is an efficient transport protocol for these applications. Forcing packets into fixed SONET time slots results in significant amounts of wasted bandwidth.

- SONET and SONET hardware are very expensive. It's estimated that Ethernet equipment costs are *70 percent less expensive* than SONET equipment.

- SONET is inherently wasteful in terms of the data transported versus the bandwidth dedicated. For example, sending 10 Mbps of Ethernet LAN traffic across a metropolitan network using SONET infrastructure requires a full DS-3 (OC-1) of bandwidth (a 51 Mbps optical channel). Because only 10 Mbps of capacity is required, 80 percent of the available capacity of the OC-1 channel is wasted! Mapping a 100 Mbps Ethernet connection to an OC-3 SONET service is less wasteful but still inefficient. In this case, 35 percent of the capacity of the OC-3 circuit is idle; 100 Mbps is delivered over a fixed 155 bandwidth pipe.

- Since it is a packet-oriented protocol, Ethernet suffers none of the limitations and inefficiencies of TDM-based equipment and transport. Ethernet is built for unpredictable traffic patterns. Furthermore, metropolitan Ethernet is perfectly matched to the standard LAN Ethernet line rates of 10 Mbps, 100 Mbps, and 1 Gbps. Some *Optical Ethernet Provisioning Platform* (OEPP) equipment provides even greater bandwidth flexibility by enabling provisioning (and billing) in

1 Mbps increments. This is the equivalent of adding a DS-1—the current "king of private lines"—in hours or days instead of weeks. Thus, customers of Ethernet service providers can subscribe to standard-rate Ethernet services or perfectly tailor their bandwidth subscription to their own needs. This rate-shaping ability is becoming a significant differentiator for Ethernet service providers (such as Telseon) and the equipment vendors that supply them (such as Extreme Networks).

In SONET networks, capacity increases require expensive systemic upgrades to the entire SONET system to include each and every node (mux) on a given ring. This is a very expensive and time-consuming proposition, and takes several months to complete most of the time. SONET provisioning also requires manual configuration of the muxes, which tends to make it inefficient in today's world of autodetection and up-and-coming technologies such as *flow-through provisioning* via *graphical user interfaces* (GUIs). These are the reasons why SONET is not optimized for a data-centric world. Table 4-2 illustrates how much less expensive Ethernet is compared to SONET when transporting 150 Mb worth of traffic.

Figure 4-4 shows the complexity of a traditional legacy SONET infrastructure in the PSTN when transporting Ethernet traffic from a business LAN. Figure 4-5 shows how much flatter this architecture can be when only GigE is used for transport in the metro network.

4.2.9 The Pros and Cons of Ethernet and SONET

To summarize, a listing of the benefits and drawbacks of each technology (as of 2002) is listed in the following sections. Like many choices in this world, each has their place, and each has their pros and cons.

Table 4-2

Comparison of 150 Mb transported over Ethernet versus traditional OC-3 SONET (costs are estimated averages charged by a traditional ILEC)

	150 Mb Ethernet (costs below are prorated)	OC-3 SONET
Monthly recurring charges	$5,540	**$7,000**
Hardware cost	$800	**$20,000**
First-year costs	$67,380	$105,250
Bandwidth costs after two years	$132,960	$168,000

Figure 4-4
Ethernet transport in
a legacy telco
architecture

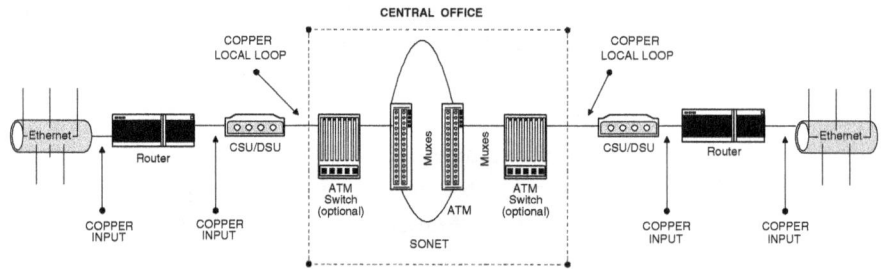

Figure 4-5
Point-to-point
Ethernet transport
through PSTN using
only Ethernet
technology (Ethernet
private line [EPL])

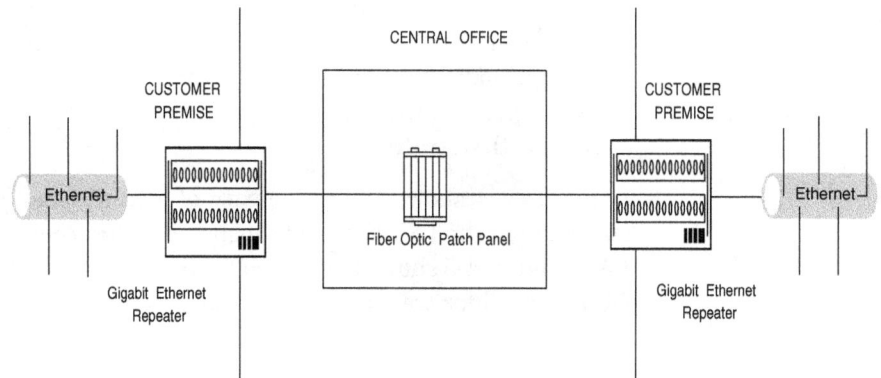

4.2.9.1 Ethernet

4.2.9.1.1 Positives

- Is scalable in increments from 1 Mbps to 1 Gbps and soon to 10 Gbps. Down the road, we'll surely see another level.

- Ethernet equipment costs are much lower than SONET equipment costs.

- Ethernet is versatile and can easily support IP services.

4.2.9.1.2 Negatives

- QoS issues exist. Ethernet isn't optimized for voice traffic.
- Ethernet isn't quickly restored in the event of a failure.
- It lacks resiliency.

Various efforts have been made to overcome these drawbacks to metro Ethernet. Several vendors such as Appian Communications, Lantern Communications, Nortel Networks, and Luminous Networks are working on very promising solutions. These companies want to tap the ILEC market now that CLECs have largely failed.

4.2.9.2 SONET

4.2.9.2.1 Positives

- Survivable, 50 ms restoration in the event of a node or link failure (99.999 percent reliability)
- Optimized for voice traffic
- Billions of dollars of deployed infrastructure

4.2.9.2.2 Negatives

- Expensive to deploy.
- Not proficient in carrying data traffic.
- Billions of dollars worth of SONET infrastructure is already deployed, and this investment has to be accounted for—it must be used until fully depreciated.
- Long provisioning cycles.
- Difficult to upgrade.
- As a narrowband TDM technology, SONET is less efficient for multiplexing high-speed, bursty data and *Motion Picture Experts Group* (MPEG) video traffic. Cell and packet-multiplexing protocols such as ATM and GigE are well known as more efficient transport systems.

4.2.10 Optical Ethernet Provisioning Platforms (OEPPs): Market Position

OEPP vendors will primarily compete against SONET vendors, particularly vendors marketing next-generation SONET products or *multiservice provisioning platforms* (MSPPs). Competing vendors in this area include

Cisco Systems (ONS 15454), Nortel Networks (OPTera Metro 3500), Fujitsu (ADX 2400), and CIENA (k2 platform).

The key advantage of SONET is that SONET ADM equipment is already installed on nearly every lit fiber ring in every incumbent carrier's metropolitan network. IDC believes that incumbents will not rip out SONET installations, but will build parallel networks when they deploy OEPP equipment instead. The ILECs may also deploy EoS on an as-needed basis, which is also sometimes known as PoS.

The emerging MSPP products enable incumbent carriers to preserve their legacy infrastructure and roll out newer Ethernet services at the same time. Although OEPP products handle Ethernet traffic better than MSPP products being sold today, it's believed that the data capabilities of the MSPP products are sufficient. At the same time, the TDM functionality of MSPPs is critical for incumbents' bread-and-butter voice revenue streams.

4.3   Fibre Channel

The interconnect between these systems and their *input/output* (I/O) devices demands a new level of performance in reliability, speed, and distance.

Fibre Channel is a highly-reliable, 1 and 2 Gbps interconnect technology that allows for concurrent communications among workstations, mainframes, servers, data storage systems, and other peripherals using *small computer systems interface* (SCSI), IP, and a wide range of other protocols to meet the needs of data centers. The standards for Fibre Channel are specified by the Fibre Channel Physical and Signaling Standard and the ANSI X3.230-1994, which is also *International Organization for Standardization* (ISO) 14165-1. Even though Fibre Channel is *compatible* with the SCSI standard because Fibre Channel is three times as fast, it has actually begun to *replace* the SCSI standard as the transmission interface between servers and "clustered" storage devices. Fibre Channel is the premier technique for *storage area networking* (SAN) and has proven its ability to deliver a new level of reliability, availability, security, and both scalable throughput and capacity. Switches, hubs, storage systems, storage devices, adapters, and management software are among the products that are on the market today that provide the ability to implement a complete Fibre Channel solution.

There are two basic types of data communications between processors and peripherals: channels and networks. A channel (as in "Fibre Channel") provides a direct or switched point-to-point connection between the communicating devices. A channel is typically hardware-intensive and transports data at high speed with low overhead. In contrast, a network is an aggregation of distributed nodes (like workstations/PCs, file servers and peripherals) with its own protocol that supports interaction among these nodes. A network has relatively high overhead because it is software-intensive and consequently slower than a channel. Networks can handle a more extensive range of tasks than channels as they operate in an environment of unanticipated connections. Channels, however, operate amongst only a few devices with predefined addresses. *Fibre Channel attempts to combine the best of these two methods of communication into a new I/O interface that meets the needs of channel users and network users alike.*

The information explosion and the need for high-performance communications for server-to-storage and server-to-server networking are the focus of much attention today. Increasingly data-intensive and high-speed networking applications have been generated by the following:

- Performance improvements in data storage technology

- Speed and capability improvements in processors and workstations

- The move to distributed architectures such as client/server

Today's data explosion presents unprecedented challenges incorporating a wide range of application requirements such as storage networking, cluster computing, database and file management, transaction processing, data warehousing, imaging, integrated audio/video, networked storage, real-time computing, collaborative projects and CAD/CAE. These are all applications that Fibre Channel can efficiently support. The capability of Fibre Channel to serve as an effective transport medium for *storage area networks* (SANs) is even more important in 2002. This is due to more companies seeking remote, geographically distributed data storage systems in the wake of the attacks on the United States on September 11, 2001.

The key benefits of a Fibre Channel implementation include the following:

- **Multiple Topologies**   Dedicated point-to-point, shared (arbitrated) loops, and scaled switched topologies meet application requirements. This offers the customer the ability to develop a storage network with configuration choices at a range of price points, levels of scalability, and availability.

- **Multiple Protocols**   Fibre Channel delivers data via SCSI, IP, VI, ESCON, and other storage and networking protocols to meet the customer needs for storage connectivity, cluster computing, and network interconnect.

▪ **Congestion Free** Fibre Channel's credit-based flow control delivers
data as fast as the destination buffer is able to receive it in order to
meet high throughput data transfers. This facilitates applications like
backup, restore, remote replication, and other business continuance
enabling capabilities.

4.3.1   Ethernet Versus Fibre Channel

Ethernet's popularity as the transport medium of choice in SANs is on the
rise, though—it's now beginning to compete with Fibre Channel as the SAN
layer 1 technology of choice. Using Ethernet end-to-end in storage networking offers a way to seamlessly extend the enterprise LAN while using
only one transport technology. Using Ethernet in SANs would also keep the
network architecture flatter, meaning less points of failure in the network,
less hardware cost (for Fibre Channel switches), and decreased network
management requirements. In general, Fibre Channel is also a more expensive technology than Ethernet.

On the other hand, storage applications are very latency sensitive. GigE
would inherently introduce more latency to a storage network, while Fibre
Channel minimizes latency. Each Ethernet frame would create transmission overhead, especially when using the iSCSI protocol, which employs IP
over Ethernet transmissions, thus requiring regular acknowledgment
transmissions. Conversely, Fibre Channel transmissions are not only hardware intensive (equating to lower overhead overall), but these transmissions are sent in huge multimegabit blocks at a time before an
acknowledgment is required. This delivers the lower latency that is
required of storage applications, which Fibre Channel is well-suited to support. Using Ethernet and IP for storage transmissions would require regularly transmitted acknowledgments, thus increasing network latency.

To confirm the benefits of using one technology over the other (Fiber
Channel or GigE), a cost/benefit analysis would have to be performed, along
with a network study to determine the impact of additional latency when
using Ethernet versus Fibre Channel. End-to-end SAN distance would play
a key part in this equation.

4.4   Frame Relay

Frame relay technology is still the most widely deployed packet-based
transport technology in place today. It has the largest embedded base in the

enterprise segment and is still growing, even after all these years. Yes, its successor technology, ATM, is also still growing, but at a slower pace overall. Compared to frame relay, ATM is a more expensive technology to deploy and manage. The equipment is more expensive. ATM is a technology that's harder to master by WAN administrators. However, that same old song applies in this situation as well. As bandwidth needs increase, frame relay's inherent bandwidth limitations will make it a passé technology by 2010. New implementations of frame relay will probably slow to a crawl into 2004 as more efficient technologies (such as GigE and 10 GigE) become increasingly available.

By around 2005, frame relay's growth will begin to level off as metro Ethernet (and possibly other new broadband transport technologies) begins to replace it for all the reasons that were stated back in Chapter 2, "Metro Area Ethernet." For example, *Ethernet virtual private networks* (EVPNs) could mimic frame relay at a much lower cost with far greater scalability. This kind of service will be much easier to implement than frame relay or ATM, and will be more scalable at affordable cost points. It will also enable enterprise customers to adapt their network service to their bandwidth needs very flexibly—a feature not inherent to frame relay. Some GigE vendors, such as Extreme Networks, say they offer similar *virtual MAN* (VMAN) and *virtual WAN* (VWAN) aggregation capabilities today, but the jury is still out as to whether this provides sufficient security for more mission-critical, high-end applications.

Like most of the other legacy transport technologies, Ethernet will compete against frame relay in the enterprise (and carrier) marketplace. Frame relay will ultimately become its own legacy technology and will only last as long as businesses can still use (or perhaps tolerate) its inherent bandwidth limitations.

4.5 Asynchronous Transfer Mode (ATM)

Like SONET/SDH, ATM is suitable for aggregating sub-OC-3 data applications. However, ATM architectures do not offer *user-to-network interface* (UNI) with speeds faster than OC-48, and they are cost prohibitive for a variety of reasons (*network-to-network interfaces* [NNI] can be implemented at OC-192 speeds). The single biggest drawback to ATM technology is the overhead cell tax of 5 bytes per 48-byte payload. This becomes costly when doing file transfer of large data volumes.

4.5.1   How Ethernet Will Beat ATM in the Enterprise Market

ATM was developed to overcome the deficiencies of narrowband TDM switches in handling bursty data traffic. It's a statistical multiplexing and switching technology developed initially by telcos, for telcos. Therefore, ATM largely ignored the higher-level protocol requirements of the Internet, which was only in its infancy when ATM's popularity began to grow in the early to mid-1990s. Competitors of the *Regional Bell Operating Companies* (RBOCs), being the commercial data network vendors, jumped on the bandwagon and formed the ATM Forum, which accelerated the development of ATM. Once again, this left the telcos and associated ITU standards committees behind. Commercial vendors aspired to deploy ATM to the desktop and ATM LAN switches. WAN vendors and telcos fell into similar traps.

The first trap was to ignore competition from the entrenched suppliers of Ethernet adapter cards (*network interface cards* [NICs]) and LAN hubs. This competition resulted in the development of 10 Mbps Ethernet switches that offered all the benefits of ATM switches without having to replace the existing Ethernet NICs fitted to millions of *personal computers* (PCs), workstations, servers, and laptops.

Key: This delayed the manufacturing of high volume levels of ATM adapter cards, which would be required to attain a price-competitive position in the marketplace. At the same time, the *Institute of Electrical and Electronics Engineers* (IEEE) Standards Committee and associated vendors continued to develop improved versions of Ethernet (such as Fast Ethernet at 100 Mbps).

The benefit of Fast Ethernet switches and NICs was that they supported the same Ethernet frame format and auto negotiation features of Ethernet (such as *Carrier Sense Multiple Access with Collision Detection* [CSMA/CD] and 10/100 PC NICs). The latter enabled existing 20 Mbps NICs and new 100 Mbps NICs to connect to the same Fast Ethernet switches, further delaying the introduction of ATM to the desktop. The IEEE Standards Committee and associated vendors continued this ongoing development strategy, resulting in the development of the GigE standard (IEEE 802.3z). Ethernet's 10, 100, 1,000 Mbps (GigE) and 10 GigE hierarchy has maintained essentially the same frame format throughout this development timeframe, which has proven to be a major factor in reducing packet-processing overheads and related costs compared to ATM switches with Ethernet interfaces.

The second trap was the complexity of managing mesh-based ATM networks and multiple services having different QoS requirements. The management protocols, algorithms, and associated software were a long time in coming, but when they finally arrived, the benefit of ATM switches was their support for rapid network reconfiguration and the QoS requirements of telephony networks.

Key: The flip side of this success was that businesses, network vendors, and telcos now experience great difficulty in retaining key staff with the skill sets required to manage ATM networks and associated switching products.

The third trap came form the very services that the ATM vendors and telcos were aiming to address: high-speed packet data and video on demand The growing demand for packet data bandwidth developed due to the evolution of the World Wide Web. However, the IP standards were developed by the *Internet Engineering Task Force* (IETF), independently of the ATM standards—to the extent that ATM proved to be an inappropriate, unnecessary, and costly overhead layer when transporting IP traffic. In many cases, Internet traffic had to be dropped off at an ATM switch, connected to an IP router, and then connected back to the ATM switch again. This was because the 5-byte ATM cell header only supported short virtual circuit identifiers (analogous to a virtual time slot number in a SONET stream). IP addresses were 16 bytes in length and required an IP router to read them and determine where the packet should go next. This additional processing step to route IP data made it clear that ATM wasn't going to have a smooth fit in the IP world, especially the IP world of the Internet.

As Ethernet works so smoothly with *Transmission Control Protocol / Internet Protocol* (TCP/IP), it was easy to see that Ethernet would dominate connecting LANs to MANs and soon to WANs.

ATM's original designers envisioned that the majority of ATM implementations would be within (over) SONET systems. However, all the same things that are making SONET transport increasingly unappealing for today's data environments are also making ATM unappealing, such as major overhead, very complex provisioning requirements, expensive equipment platforms, and long provisioning cycles. The biggest functional drawback to ATM technology is that it is very expensive due to its inherent complexity. This higher expense never allowed the prices for ATM technology (equipment) to come down over time because ATM was never deployed in great enough numbers.

4.6 DS-1 Private Lines

DS-1 private lines are point-to-point circuits used to connect two or more locations together. These locations can be across a town, state, or country.

Key: No other type of transport medium is more widely deployed in the United States than the DS-1 circuit, which is also known as the T1.

On the revenue side, the DS-1 circuit is one of the RBOCs' true bread-and-butter revenue streams, with millions of circuits installed throughout the country equating to billions of dollars in annual revenue. Since the late 1980s, it has been the transport medium of choice and has served corporate America very reliably.

But Ethernet is starting to edge out the T1 as the transport medium of choice. Why? It has exponentially more bandwidth—up to 1,000 times more bandwidth—for prices that aren't exponentially higher, requirements for more bandwidth, and seamless extension of the LAN. However, many companies in American business are still clinging to the DS-1 circuit, and they believe it's all they need right now.

As the power of the Internet and e-commerce increase visibly while bandwidth needs grow, more businesses will see the obvious benefit to migrating their private-line technologies to GigE, especially for metro area connectivity (such as Internet access, remote office connectivity, and so on).

In the largest businesses surveyed by IDC (10,000+ employees) in a 2000 study, 79 percent of the respondents used T1 speeds and lower for their WAN backbone traffic. Only 1.3 percent of the respondents employed OC-*n* (optical) WAN backbone services. However, 2000 is the year when carriers (of all types) began offering GigE en masse. It offered cost-effective, easy-to-upgrade DS-1 MAN and WAN backbones.

The dominance of DS-1, even within the largest corporations in the United States, suggests that end users may not be rapidly migrating to gigabit speeds in the MAN. However, it is important to note that the IDC survey referred to previously was conducted in 2000. This is the very same year when the ELECs (Yipes, Telseon, and Cogent) began to penetrate the metro market with their pure-play Ethernet service offerings. Most industry analysts agree that 2000 was the year when Ethernet began its heavy push in the metro to replace services such as DS-1 circuits. What's the point? Everyone knows that change can be measured in months in the information and telecom industry, not necessarily years. The year 2000 is ancient history in this field. Many of the corporations surveyed may not have begun to apply the Internet and other packet-data technologies to their applications or e-commerce platforms in earnest. The network status and bandwidth requirements of the companies polled could have changed dramatically since 2000.

The T1 is certainly not going away anytime soon, but metro Ethernet will continue to become popular. Medium-sized and small businesses may eventually get on the GigE bandwagon as prices fall and marketing efforts are eventually targeted toward them.

Key: If the average price of T1s today is around $300 to $500 per month, and BLEC/ELEC Cogent is offering 1,000 times the bandwidth of a DS-1 circuit for $1,000 per month (and they can sustain that pricing model), it will only be a matter of time until more businesses choose pure Ethernet over circuits like the DS-1. It just makes simple economic and good business sense.

The DS-1 circuit, like its companion technologies such as frame relay, will reach the peak of its product life cycle and begin to decline in use by 2005. New installations will probably cease by 2006.

4.7   DS-3 Private Lines

DS-3 circuits will also decline for all the same reasons that DS-1 circuits will decline in popularity and use between 2005 and 2010. Additionally, DS-3 circuits will decline because compared to Ethernet offerings, DS-3 circuits are priced at similar, or even higher, rates. If a company can obtain 1,000 Mb of bandwidth for around $5,000 a month or pay $4,000 dollars a month for just 45 Mb of bandwidth (roughly 20 times less capacity), what will they do? It's a no brainer. GigE could easily displace the DS-3 circuit for those enterprise customers who are savvy enough to know they can get much more bang for their proverbial buck with GigE versus a DS-3 circuit.

4.8   Private Fiber

Another networking option that is competing with metro GigE is private fiber. Private fiber is defined as an independent entity—usually a business (with deep pockets) actually managing the build of their own private fiber network. They lease or trench their own fiber and put electronics on the end of it.

Constructing a private fiber network is a huge undertaking. It requires having someone manage the entire project, from obtaining right-of-way permits to trenching the fiber in the ground to actually lighting the fiber and putting the network into production. Permits would also have to be obtained from every municipality that the fiber would run through because all the streets would have to be dug up.

Key: Depending on the area, it could cost anywhere from $300,000 per mile to $1,000,000 per mile to trench fiber into the ground.

A trenching contractor would also have to be contracted. Once the fiber is laid, which could take months, the terminating equipment would need to be purchased. In most cases, this would be SONET ADMs and/or WDM systems.

Once the fiber is laid end to end, a contractor would have to be signed up to light the fiber using the terminating equipment. The next step would be to manage the migration of the company's traffic from whatever platform they're currently using to connect the two locations together to the new private fiber-based platform. This migration will usually be away from some type of leased transport service from the telco or a CLEC.

The final critical element in this private fiber equation would be the need to obtain personnel to manage the new equipment/network. This might entail having to hire new, highly skilled network technicians. What's more, if a break in the fiber cable occurs, the company alone is responsible for determining where the break is and would then have to manage the repair process as well. This would involve inking a contract with a business that specializes in this area as the fiber was being trenched end to end. One advantage to implementing private fiber is that if enough fiber is laid (which is usually the case), the entity that laid the private fiber could even lease part of it (dark fiber) to other entities. As a matter of fact, many villages and municipalities are getting into this game. They're actually offering dark fiber to businesses in their jurisdictions—it ends up becoming a revenue stream for the city, similar to leasing space on the top of water tanks to wireless carriers for their base station antennas.

The downsides to implementing a private fiber solution are as follows:

- The company is effectively building their own optical fiber network. This is an extremely expensive and time-consuming undertaking.
- Along with the cost of trenching the fiber, if SONET equipment is used at each end of the fiber, this alone could cost hundreds of thousands of dollars to purchase.
- The deployment of a private fiber network is a project. That means that additional funds will be required just to manage the project, and in the end, a decision will have to be made regarding the management of the network. Should it be outsourced, or should full-time employees (new, additional full-time employees) be hired to manage the network?

In the end, it is always a better proposition to pay a monthly fee to lease optical services in the metro.

Complementary Technologies and Protocols

The real definition of the metro optical network market includes technologies such as *Gigabit Ethernet* (GigE) (naturally), traditional *Synchronous Optical Network* (SONET), next-generation SONET, 10 GigE, Fibre Channel, and *wavelength division multiplexing* (WDM) technology. Cahners In-Stat/MDR expects this market to grow from a revenue base of $13 billion in 2001 to $23.6 billion in 2005, making it one of the fastest growing segments in the telecommunications industry. Cahners reports that from 2002 to 2003, legacy SONET/*Synchronous Digital Hierarchy* (SDH) networks will still be expanded and upgraded as carriers weather the current economic climate. However, newly constructed networks that are designed specifically for data-centric traffic and based on next-generation optical network systems will offer the best hope for future profitability. Ultimately, these networks will lower operating costs, improve packet-handling efficiency, and position network operators to capture revenue from new services. The long-term forecast for U.S. and international metro optical markets remains very positive in spite of the telecom meltdown of 2001 and 2002.

5.1   Ethernet over SONET (EoS)

"Every once in a while a great new architecture comes along that, if you could start over from scratch, you'd use the new architecture. That's what Ethernet is today," says Karen Barton, vice president of marketing for Appian Communications, a *metropolitan area network* (MAN) Ethernet equipment startup. "Unfortunately, with the existing base of SONET, it's unreasonable to think about its wholesale replacement. But, because SONET is so widely deployed, you can build on it."

SONET/SDH technology is globally deployed and still growing. An estimated $8 billion was spent in 2001 on equipment for metro areas alone. The advantage of SONET is still found in a highly resilient, fully managed network foundation that has back-office systems ready to be used across carrier networks. *Ethernet over SONET* (EoS) describes the mapping of Ethernet frames into SONET payloads based on the *International Telecommunications Union* (ITU) X.86 and T1x1.5 *Generic Framing Procedure* (GFP) standards. EoS effectively transforms a portion of the SONET network into an invisible tunnel between *local area networks* (LANs) to provide *transparent LAN* (TLAN) service. Instead of creating another network overlay, many carriers are leveraging a proven infrastructure to create carrier-class EoS/SDH services featuring the following:

- 50 ms restoration/recovery
- Metro and wide area service scope and scale
- Easy integration with back-office systems
- End-to-end service-level management

However, as noted in Chapter 4, "Competing Technologies," it's well understood that the technology was designed for a different time and set of network services: voice communications. SONET is a circuit-optimized, *time division multiplexing* (TDM) solution that is well suited for transporting voice and private-line data traffic. But again, today's traffic and revenue growth is coming from packet applications, not voice. These applications are characterized by bursty traffic patterns, diverse connectivity requirements, and a demand for bandwidth that can fluctuate widely based on the application, time of day, season, and location.

Key: An EoS architecture that combines Ethernet economics, packet-switching flexibility, and SONET resilience offers a compelling cost-of-ownership strategy for carriers. An important benefit of this architecture is that it is extendable beyond the metro area to include any wide area geography.

Providers who can leverage their existing SONET infrastructure will also be able to more cost effectively deploy new packet services without building a parallel network infrastructure or being limited to a regional scope. One downside to this approach is that service providers will pay an equipment-cost penalty as SONET networks must be overbuilt to accommodate growth in packet traffic. A better solution, according to the *Resilient Packet Ring* (RPR) proponents, would be a new packet *Medium Access Control* (MAC) that uses rings efficiently, but that also exhibits the resilience and *quality of service* (QoS) of SONET. The better solution, of course, is RPR.

Although SONET scalability is also a benefit of this approach (OC3/12/48/192), the scalability is very expensive, clumsy, difficult to provision, and burdensome compared to Ethernet's near-instant, 1-Mb-at-a-time capability to scale as of 2002.

Key: SONET is a transport-only, layer 1 technology that relies on higher-level protocols and switching systems to establish logical connectivity. This enables SONET to support multiple services including TDM, *Asynchronous Transfer Mode* (ATM), frame relay, and *Internet Protocol* (IP)-based traffic.

5.1.1 Data over SONET

Neither SONET approach—*Bidirectional Line-Switched Ring* (BLSR) or *Unidirectional Path Switched Ring* (UPSR)—is optimized for data traffic and new emerging Ethernet services. The way around this limitation for many carriers is to embed packet efficiency into the existing SONET/SDH infrastructure. This would result in a much lower cost of ownership (per megabit) by aggregating traffic from many customers onto shared SONET/SDH paths. Some solutions, such as Appian's *Optical Services Activation Platform*™ (OSAP™), add QoS and layer 2 switching to the mix in their equipment, thus enabling highly efficient services guaranteed by a *service level agreement* (SLA) that achieve mesh flexibility and economics over already-deployed SONET/SDH rings.

IDC believes that incumbent service providers, who represent the vast majority of equipment spending, will leverage their existing SONET infrastructures to the maximum effect. And why not? It makes economic sense considering the billions of dollars they've spent on the technology. This bodes particularly well for equipment vendors building Ethernet functionality onto SONET equipment (such as Cisco, Nortel, CIENA, and Fujitsu).

Most service providers are not offering packetized voice currently, but it's either in development or included in their product roadmaps. QoS (latency) and redundancy are key issues. Service providers like their existing SONET rings, and for good reason. The rings reach more customers with less fiber than other network topologies and require fewer switch ports at busy hub sites. However, it's clear that SONET wastes bandwidth when carrying packet data traffic. *Packet over SONET* (PoS) offers a partial solution, but only for point-to-point links.

Key: The carrier-class resilience of SONET underlies all higher-level network services, ensuring that transport network failures are healed before the higher-layer protocols or network users detect a fault.

Carriers stand to gain some benefits with a network solution that adds packet-switching efficiency and flexibility to their existing SONET/SDH networks. The ability to share previously dedicated SONET/SDH time slots across multiple customers is central to providing efficiency in next-generation SONET systems. This is also a prerequisite to protect carrier-grade security and service-level guarantees.

Four key service parameters are required to deliver a next-generation packet-optimized SONET/SDH solution: secure traffic aggregation, explicit rate QoS, a distributed switching architecture, and RPR protection.

5.1.1.1   Secure Traffic Segregation

Secure traffic segregation ensures that traffic to and from each individual customer port is completely isolated from the traffic of all other customers. To bring the efficiency of packet switching to a SONET/SDH network, the first requirement is to share what has historically been a dedicated, single-service (one customer) time slot among multiple users. The ability to securely isolate one customer's traffic from another's is primary to achieving this goal. In the most secure example, an individual *Ethernet private line* (EPL) service can be mapped to a dedicated (SONET) time slot. Although this approach does not leverage shared path economics, it still provides a lower cost of ownership to both the customer and carrier. This is because special-purpose *wide area network* (WAN) access equipment is replaced with Ethernet equipment that can be remotely provisioned to new service rates with much greater granularity. It also offers the benefit of *physical* (PHY) layer security with the service-level guarantees that are typical of today's private-line services such as DS-1.

Key: The ultimate effect of the design described previously is that it is a much cheaper per connection as many customers securely share network resources. This requires that the traffic for each service be isolated via standard (*virtual LAN* [VLAN]) *tagging* or *labeling* mechanisms (802.1q or *multiprotocol label switching* [MPLS]).

5.1.1.1.1   Tagging

Early EoS implementations used standard *Institute of Electrical and Electronics Engineers* (IEEE) 802.1q VLAN tags to separate one customer's traffic from another's. Unfortunately, this approach does not scale well and is inherently insecure. For example, the IEEE 802.1q standard defines a 12-bit tag that is specific to a physical port interface, but it limits the maximum number of virtual connections (or isolated customer flows) to 4,096. Proprietary solutions have been developed to resolve these inefficiencies, but because they're proprietary, they don't have the capability to extend across the wide area. This inhibits the appeal of Ethernet in the WAN, but it's overcome by the other benefits that have been listed.

Security is another major concern. It's possible for any switch on a network to join a VLAN by learning the tag assigned to the port. Since a VLAN is a broadcast service, once a group has been dynamically added, it can reach any other node on the network. In effect, this puts every node at risk. Since each layer 3 switch makes an independent forwarding decision, end-to-end service guarantees are not feasible. Traffic is typically rate limited versus shaped, with the indiscriminate clipping or dropping of packets. This type of transport instability is unacceptable for many applications.

Key: The MPLS standards work currently under way proposes a direct link between Ethernet and MPLS, which means wrapping Ethernet frames in MPLS labels to enable the separation and prioritization of each customer's traffic. In this context, the key values of MPLS are

- The ability to define an end-to-end connection similar to a *permanent virtual circuit* (PVC)
- A labeling scheme for defining connection service levels
- A labeling scheme that scales across a large community of users

As MPLS is more fully deployed in the core of carrier networks, it is also envisioned that signaling information contained within the service header can be used to achieve end-to-end bandwidth guarantees and service-level monitoring.

5.1.1.2 Explicit Rate QoS Explicit rate QoS can support shared SONET/SDH path network economics while delivering services with explicit rate guarantees, flexible burst rate options, and policy-managed bandwidth management. QoS is a critical requirement to bring packet data economics to legacy SONET networks. Although the objective is to securely share a SONET payload across multiple users or services, the risk is that traffic bursts from one service can rob the others of required bandwidth. A sophisticated QoS and prioritization scheme is needed to deliver packet services with the service-level guarantees expected by end users, especially if carriers will pay penalties for not meeting SLA levels.

QoS is also required to deliver multiple services via a single subscriber interface. For instance, an Ethernet subscriber interface that provides access to an EPL, Internet access, and frame relay corporate *virtual private network* (VPN) service must ensure that each service is given its committed bandwidth and service level.

Key: This multiservice Ethernet model greatly reduces complexity and cost for the customer. It also enables per-customer revenue to grow with a minimal investment in new infrastructure by carriers.

Like ATM transport, bandwidth in these new systems should be dynamically managed using QoS mechanisms that control the guaranteed minimum and maximum burst rate characteristics of each individual service.

Class of service (CoS) prioritization should also be employed based on Diff-Serv, MPLS, or 802.1p specification. This will ensure that higher priority traffic is forwarded first. This type of network architecture would ensure that network bandwidth is efficiently and fully utilized.

5.1.1.3 A Distributed Switching Architecture A distributed switching architecture is needed for EoS that provides the ability to switch any single service on a subscriber interface to any SONET/SDH path in the network—in both point-to-point and multipoint configurations. A distributed packet architecture that enables traffic from multiple customers to share a common payload or path through the network would be ideal. Statistical multiplexing should also be employed to ensure that bandwidth on all pipes and routes is fully optimized.

Key: This configuration could be thought of as a *secure Ethernet cross-connect*, distinguishing itself by enabling any single service on a given subscriber port to be mapped to any SONET/SDH payload while retaining explicit bit rate guarantees. Once traffic is in a path, it can be switched to another path without requiring SONET/SDH level reconfiguration.

Allowances for multiple ingress and egress points on a particular path should also be included in this design. This would provide even greater network-level bandwidth efficiency while also enabling a broader range of point-to-multipoint services. Different services and customers could dynamically use any part of a path through the network instead of dedicating many separate point-to-point paths for all of them. This would optimize the heck out of the network!

5.1.1.4 RPR Protection RPR protection is needed to provide a highly efficient and resilient mechanism for packet services to share SONET/SDH paths in widely deployed ring networks. Here, Ethernet provides a simple layer 2 entry point to services that are far more granular and scalable than traditional TDM services. IP directly over SONET is expensive and difficult to manage when the number of links becomes large. EoS, on the other hand, adds a switching capability that can route a number of point-to-point links (in other words, a full mesh may not be required). EoS enables LANs, MANs, and WANs to be combined to form end-to-end connections, thereby reducing the need for format and protocol conversions within the network. By replacing circuit-oriented WAN access technologies (such as ATM and frame relay) with an Ethernet demarc, carriers can provide simple, software-provisioned access to point-to-point or multipoint services that include EPLs, layer 2 VPNs, and TLANs. They'd also have the opportunity to greatly reduce network transport costs by aggregating Ethernet traffic from multiple customers to shared SONET/SDH paths that terminate directly at a particular carrier *point of presence* (POP).

A variation on the previous scenarios is to transport Ethernet over a WDM-based PHY layer, with or without a thin SONET interface. Here, a small subset of the SONET header would be used—the majority of the overhead is eliminated. This solution avoids most of the complexities of SONET TDM functions, the stringent SONET PHY layer specifications, and the need for a separate SONET *element management system* (EMS). The key advantage here is that the overheads of ATM and SONET can be eliminated. Both 1 GigE and 10 GigE are more affordable, practical, and simpler than ATM, the major alternative for high-speed WANs.

5.1.2 Virtual Concatenation: The Best Option for EoS?

In December 2001, a company known as PMC Sierra, Inc. announced the development of a silicon chip that sandwiches two GigE channels into an OC-48 (2.5 Gbps) SONET signal. The chip uses virtual concatenation and GFP standards that were finalized in late 2001. Both standards have been developed to offer more bandwidth-efficient ways of packing Ethernet traffic into a SONET/SDH transport network.

Key: GFP makes it possible to transport any protocol over SONET/SDH, including GigE, Fibre Channel, *enterprise system connectivity* (ESCON), and *digital video broadcast* (DVB). The key benefit to using GFP is that there is no degradation of service and no waste of bandwidth.

To understand why virtual concatenation is so valuable involves looking at how data is transported over existing networks. Simply put, SONET channels are the wrong size for carrying Ethernet traffic. In concatenated SONET, the channel is simply treated as a fat pipe. For example, putting a single GigE data stream (1 Gbps) into an OC-48c (2.5 Gbps) channel wastes 58 percent of the bandwidth of the SONET channel. Table 5-1 illustrates this point by showing how much bandwidth is saved versus wasted when using virtual concatenation.

Table 5-1

SONET virtual concatenation bandwidth utilization versus waste

Service	Bit Rate	Bandwidth Utilization without VC	Bandwidth Utilization with VC
Ethernet	10 Mbps	STS-1 (20%)	VT1.5-7v (89%)
Fast Ethernet	100 Mbps	STS-3c (67%)	STS-1-2v (100%)
GigE	1 Gbps	STS-48c (42%)	STS-3c-7v (95%)
Low-speed ATM	25 Mbps	STS-1 (50%)	VT1.5-16v (98%)
Fibre Channel	200 Mbps	STS-12c (33%)	STS-1-4v (100%)
Fibre Channel	1 Gbps	STS-48c (42%)	STS-3c-7v (95%)

An improvement on this scheme is *channelized SONET*, which enables carriers to carve up capacity in an OC-48 link into units of STS-1 (51.4 Mbps), STS-3 (155 Mbps), or STS-12 (622 Mbps) channels. However, it's just not possible to mix and match these units because there's no guarantee that the same combination of units will be available at both ends of the network, if at all.

Key: Enter virtual concatenation. In a nutshell, it's job is to *right-size* SONET channels so more efficient provisioning of bandwidth can occur, namely for Ethernet (packet) traffic. Notice that more efficient isn't necessarily ideal.

Transporting EoS using virtual concatenation could be a better option than native Ethernet networks for established carriers. Carriers wouldn't need a forklift upgrade to their existing networks, which are predominantly built on channelized SONET. To support virtual concatenation, a carrier only has to upgrade the equipment on the ends of the connection. Further, SONET offers features that native Ethernet currently doesn't have such as bandwidth guarantees and protection in the event of a network failure.

PMC is not the only chip maker that thinks support of EoS is a smart move. In October 2001, Agere Systems also unveiled what it dubbed an *add/drop multiplexer (ADM) on a chip*, which supports both virtual concatenation and GFP standards like PMC. However, the design of the chips from Agere and PMC are quite different. Agere's chip integrates a pointer processor with an STS-1 granularity cross-connect feature. Conversely, PMC's chip doesn't include the STS-1 cross-connect, but it does have the capability to interface directly to an optical module on one side and a backplane on the other side, if desired. "This is one of the highest-level integration devices we've ever produced," according to a source at PMC.

It's also worth noting that system developers such as Sycamore Networks have had equipment for more than a year that crams two GigE channels into a 2.5 Gbps wavelength. Bear in mind even this device wastes 500 Mb of bandwidth, the equivalent of over 300 DS-1 circuits! However, Sycamore had to develop its own type of silicon to achieve this feat. The off-the-shelf chips now being offered by Agere and PMC make it easier for other system vendors to catch up.

5.1.3 Summary: EoS

In all cases, the economic benefits of Ethernet play out in lower cost and complexity for both carriers and customers. Unlike today's hard-wired TDM circuit interfaces, an Ethernet interface can be remotely provisioned to adjust bandwidth or change service parameters. Customers no longer need to schedule truck rolls or circuit provisioning upgrades because the inherent flexibility of Ethernet supports connections that scale from 64 Kbps to 10 Gbps. Now that's flexible!

5.2 Transmission Control Protocol/Internet Protocol (TCP/IP)

Ethernet is by far the most successful networking technology ever. Ninety-nine percent of all *Transmission Control Protocol/Internet Protocol* (TCP/IP) packets (including Internet traffic) traverse at least one Ethernet somewhere and more likely five or six Ethernet LANs.

Computers attached to an Ethernet network can send application data to one another using higher-level protocol software, such as the TCP/IP protocol suite, which serves as the foundation of the Internet. The higher-level protocol packets are carried between computers in the data field of Ethernet frames.

5.2.1   The IP-Optimized Network

Next-generation MANs demand an architecture optimized for differentiated (multimedia) IP-based services. This requirement reflects the slow but steady move toward convergence. Simultaneously, these next-generation MANs must provide support for today's legacy voice services, which continue to provide a significant source of revenue for service providers. These objectives can only be achieved through a packet-switched network that provides end-to-end service guarantees in terms of latency, jitter, bandwidth, and packet loss. Along with Ethernet, technologies under development will provide these guarantees (such as MPLS and RPR). Figure 5-1 illustrates how an IP over GigE network can streamline the cost and complexity of today's legacy networks.

Figure 5-1
How IP over GigE can simplify data networks

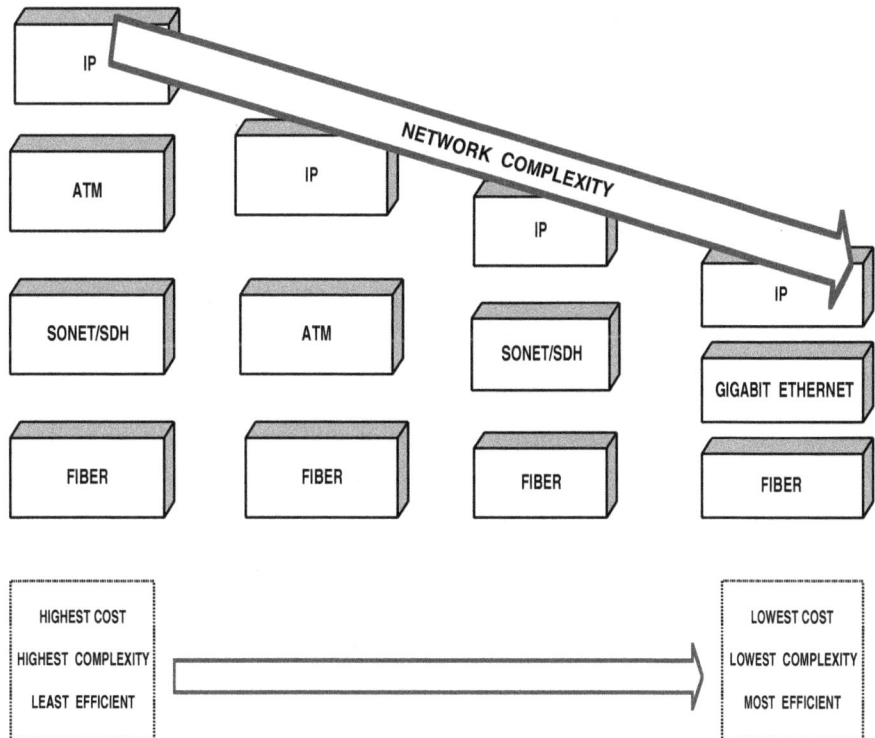

5.3 Wavelength Division Multiplexing (WDM)

By the mid-1990s, businesses discovered that SONET and ATM were ineffective options for the higher-speed data protocols that they increasingly sought to adopt, such as IP, ESCON, Fibre Channel, and GigE. Since 2000, WDM has emerged as a viable option to support these enterprise applications.

WDM is a technology that enables multiple optical signals to be transmitted (multiplexed) over a single fiber using the 1,300 or 1,500 nm wavelength windows (optical frequencies). These frequencies are light-based and resemble numerous, minute divisions of the colors that comprise white light —the colors of the rainbow. As of 2002, each wavelength is capable of transmitting up to 9.953 Gbps of traffic (one OC-192 or a native 10 GigE data stream), typically in a point-to-point configuration. Two types of WDM are available from equipment makers today: *dense wavelength division multiplexing* (DWDM) and *coarse wavelength division multiplexing* (CWDM). GigE is equally compatible with both.

5.3.1 DWDM

DWDM is a technology that puts data from different sources together on an optical fiber, with each signal carried at the same time on its own separate optical (light) wavelength. Using DWDM, up to 160 (unprotected) separate wavelengths or channels of data can be multiplexed into a light stream transmitted on a single optical fiber. Each of these channels could carry 1 GigE of traffic. Since each channel is demultiplexed at the end of the transmission back into the original source, different data formats being transmitted at different data rates can be transmitted together. Specifically, Internet (IP), SONET, ATM, or Ethernet data can all be traveling at the same time within the optical fiber, each on their own wavelength.

DWDM systems are widely deployed in long-haul networks worldwide. DWDM promises to solve the fiber-exhaust problem and is expected to be the central technology in the all-optical networks of the future. Naturally, the capacity of this technology is bound to increase depending on the pace of development. The sky is probably the limit.

5.3.2 CWDM

CWDM works on the same principle as DWDM, but the spacing of the wavelengths in CWDM systems (the frequency width assigned to each optical channel) is much wider in CWDM systems than in DWDM systems. This is why CWDM systems always have lower capacities than DWDM systems. CWDM systems are also less expensive than DWDM systems because the lasers in CWDM equipment do not require cooling as they do in DWDM systems. This dramatically reduces the cost of CWDM equipment. Today's CWDM systems can handle capacities of 64 unprotected wavelengths or 32 protected wavelengths.

The following are key elements and functions of a WDM:

- Lasers are used as transmitters, one for each wavelength.
- A WDM mux combines multiple optical signals for transmission over fiber strands.
- Optical amps are used to pump up the signal power and compensate for system losses as required. DWDM networks also permit longer unrepeatered transmission distances compared to TDM networks that have the same total bandwidth.
- An optical demux separates the received signal into its multiple wavelengths for delivery to receivers that then convert the signal back to electrical form (*optical-electrical-optical* [OEO] conversion).
- Optical signals can be added and removed from WDM systems using optical ADMs, which groom and split the optical signals along the transmission path of the network.

The next generation of WDM will support *light paths* (optical wavelengths), which can be established and released using a dynamic management system rather than through static configurations. This is known as *lambdas on demand service*.

WDM provides an alternative to SONET-based technologies that enables a better match of bandwidth supply to user demands while also making provisioning much more flexible and dynamic. WDM also eliminates some of the legacy baggage of SONET technology such as considerable protocol overhead and high deployment costs.

Key: Because WDM operates at layer 1 of the OSI model (the PHY layer), it can carry any signaling protocol and can even carry different protocol traffic on each individual wavelength. This is known as *protocol transparency*, or *protocol agnosticism* in the context of datacom transport. Because DWDM solutions are not bound by the bandwidth hierarchy defined for SONET (that is, OC-48 at 2.488 Mb or OC-192 at 9.953 Mb), they can be adapted for use directly with Ethernet. Flexible capacity management of this nature will enable carriers to offer bandwidth on demand once WDM systems are widely deployed in carrier networks (possibly around 2004 and 2005).

Table 5-2 shows the evolution of WDM technology since the 1980s. Note that capacity has increased as channel spacing has narrowed over time—a logical progression.

WDM technology provides a number of advantages for metro network environments. Clearly the most important advantage is its capability to optimize existing fiber installations. In the majority of cases, adding new (WDM) terminating equipment to optical networks to increase the number of channels is much less expensive than physically laying new fiber. Over time, this will also bring down the costs of WDM gear as more of it is sold and it becomes cheaper to manufacture. This reflects a key law of economics: The more of something that's produced and sold, the cheaper it becomes to make—and these savings are usually passed on to the buyers. DWDM provides the most flexible backbone network architecture since each wavelength can transport different protocols and signal formats.

Metro DWDM systems are now available at a cost of about $25,000 per wavelength ($12,500 per end). In fact, many carriers are realizing a full return on their investments in less than one year because both the carriers

Table 5-2

Evolution of WDM technology since 1980

2000	80+ (unprotected) channels 0.8 nm spacing WDM, integrated systems with network management and add/drop functions
1990s	2 to 4 channels 3 to 5 nm spacing Passive WDM components/parts
1980s	2 Channels *Wideband WDM* (WWDM) 1,310 and 1,550 nm

and enterprises recognize the new relationships for what they clearly are: win-win scenarios. The affordability of the networking solutions that make DWDM services possible has enabled carriers to begin targeting market segments they could not previously support. Because the cost to the carriers is relatively low, they are pricing the services aggressively. Because the cost for the enterprises is low, demand is significantly increasing.

The introduction of WDM enables many increases in bandwidth on one strand of fiber. In the carrier backbone and an increasing part of the metro core, WDM muxes are starting to replace traditional SONET muxes. This creates a plentiful supply of cost-effective bandwidth in the metro core. Combining transmission capacity with rapid service provisioning and dynamic wavelength management could equal reduced cost and deployment time for MAN infrastructures.

DWDM systems use higher-level protocols such as ATM (layer 2) or TCP/IP (layer 3) for logical path restoration. Since restoration requires the logical rerouting of traffic through the network, it is much slower than the 50 ms protection switching performed by SONET rings. To address this drawback, next-generation DWDM optical switches will offer ring protection based on full or shared wavelength protection paths that are similar to the UPSR and BPSR protection schemes employed by today's SONET networks.

Next-generation metro networks will run Ethernet and other traffic over DWDM without the use of SONET muxing. This is a powerful incentive to migrate toward a pure DWDM platform.

5.3.3   Ethernet over DWDM

The rapid growth in Internet traffic compared to traditional voice telephony traffic has caused many network architects to stand back and reconsider the wisdom of designing broadband data-centric networks on top of narrowband telephony networks (SONET).

At the same time, the emergence of new DWDM technologies and products have provided incumbent carriers (such as *Regional Bell Operating Companies* [RBOCs]), new carriers (such as *Competitive Local Exchange Carriers* [CLECs]), and carrier's carriers with greater flexibility in selecting new protocols and architectures to match their various customer service requirements. For the carriers, DWDM rings will become the preferred architecture for fault-tolerant metro and access network applications. For business applications, point-to-point DWDM networks can be a popular solution that minimizes fiber lease costs.

Vendors are now building metro DWDM products with Ethernet interfaces, enabling optical Ethernet traffic to run directly over WDM wavelengths. This equipment will play an increasingly important role in carrier infrastructures for Ethernet services. This architecture underscores not only how WDM is complementary to Ethernet, but also how both technologies are serving to dramatically flatten network topologies. Refer to Figure 5-1, where the network is flattened, and imagine a DWDM layer underneath the FIBER box to understand where DWDM exists in the network.

By establishing a DWDM sublayer under layer 1 (dark fiber), the metro core network will have access to virtually unlimited bandwidth. The SONET core network can be capped, handling slower-growing, TDM-based traffic. Data-centric IP traffic could increasingly be routed from the access network directly to a DWDM wavelength. Over an extended period of time, TDM traffic can be migrated to the DWDM layer as well, eliminating the costly, complex SONET metro core network altogether. By not imposing additional, arbitrary framing structures, this new convergence-optimized network architecture allows for flexible allocation of bandwidth by the wavelength and more efficient utilization of all available bandwidth.

Metro DWDM equipment will typically reside in the metro core network, accepting optical Ethernet traffic from *Optical Ethernet Provisioning Platforms* (OEPPs) at the edge of the network. In this respect, DWDM will compete more directly against SONET in the metro. Besides creating more efficiency and flexibility across the MAN, the direct deployment of Ethernet in a WDM transport infrastructure greatly enhances a carrier's ability to extend broadband connectivity over the last mile to customers.

One new architecture option is the direct provisioning of multiple GigE channels over WDM (fiber) to create a scalable, robust, and data-optimized optical ring. As a natural extension to GigE, 10 GigE will also provide an inherently scalable means to efficiently and flexibly aggregate native Ethernet traffic across high-speed optical MANs. Figure 5-2 shows how easily an Ethernet/IP network can be integrated over a WDM infrastructure. Integrating WDM directly into a layer 2/3 switch chassis boosts capacity, simplifies network deployment and management, and reduces support costs.

The key point here is that for metro and access network applications, GigE will be a major driver in determining the maximum unit of capacity that is accessible in the time domain, with all subsequent increases in circuit capacity provided in the wavelength (optical) domain. Using this approach, 1 to 100 GigE networks are readily supported and are easily accessed using simple, low-cost optical ADM and 24-port 10/100 Ethernet to GigE switching products.

Figure 5-2
A converged switch in the new architecture can collapse into a single system all of the functionality that required a WDM mux, a SONET ADM, and ATM switches under the legacy SONET paradigm.

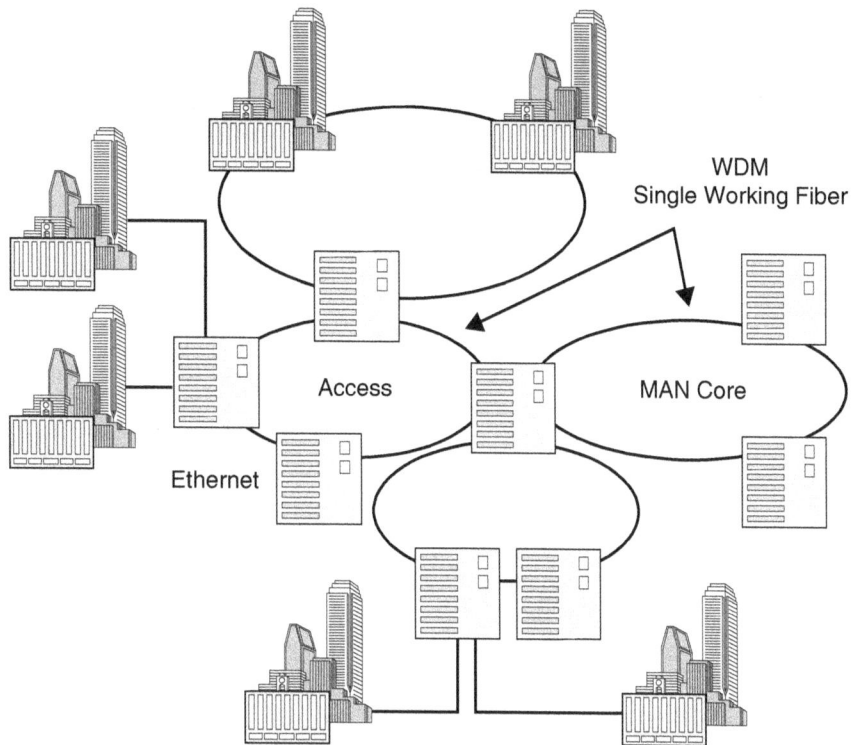

Figure 5-2
A converged switch in the new architecture can collapse into a single system all of the functionality that required a WDM mux, a SONET ADM, and ATM switches under the legacy SONET paradigm.

5.4 Next-Generation SONET

The optical marketplace has undergone a revolution since the late 1990s, and today optical solutions garner much of the limelight for technical innovations. However, for every GigE, *passive optical network* (PON), or all-optical startup, solutions are available that build on existing legacy SONET networks, promising more flexibility and ease of use for carriers and enterprise managers.

A new class of SONET network equipment is being developed to address major metropolitan network requirements and existing SONET deficiencies. These next-generation SONET products offer increased capacity (up to OC-192) and also support multiple services such as Ethernet, ATM, IP, and TDM. They typically combine the functionality of SONET ADMs and standalone *Digital Access Cross-connect Systems* (DACS). Data switching and DWDM transport capabilities are sometimes also included.

Key: These systems use statistical multiplexing rather than fixed TDM time slots to improve packet-handling efficiency. They also permit over-subscription.

Advanced silicon technologies are being used in these boxes to significantly reduce unit size and cost. The multiservice, multifunction design of next-generation SONET boxes is targeted at both metro core and (metro) access network applications.

Many solutions are available that use SONET interfaces today for WAN connections, including routers with PoS interfaces and ATM switches implementing SONET back-haul links. However, equipment that just inserts traffic into a SONET frame without adhering to basic principles and key SONET attributes, such as traffic engineering, multiplexing, and traffic management, cannot really be named a next-generation SONET solution. The need for next-generation SONET stems from the necessary upgrades and expansion of existing SONET networks.

The next-generation SONET market in North America is forecasted to grow to $6.3 billion by 2004. This growth is being driven mainly by traditional carriers that are evolving their SONET-based networks along a careful migration path to next-generation SONET and other new optical technologies.

Several features distinguish next-generation SONET from other technologies:

- Total backward compatibility with legacy SONET systems
- Greater port density in smaller physical footprints
- Layer 2 functionality with support for Ethernet
- Embedded intelligence distributed to all the *network elements* (NEs)
- Faster provisioning through intelligent software
- Full integration with common *operational support systems* (OSSs)

Next-generation SONET systems are typically installed in the basement or demarc area of the customer premises. The customer sends Ethernet traffic to the SONET box over a simple 10, 100, 1,000, and soon 10,000 Mbps Ethernet LAN connection. The next-generation SONET mux aggregates Ethernet streams from multiple customers and maps them onto an OC-12 (622 Mbps) or OC-48 (2.488 Gbps) payload for transport over the embedded legacy SONET network. (Ethernet is encapsulated into SONET.)

Next-generation SONET systems have the advantage of backward compatibility with legacy SONET systems. Although these systems offer significant improvement over legacy equipment, they still remain fixed systems—in other words, equipment at all working and protection ring nodes must be upgraded as necessary, which mandates a large up-front investment. Figure 5-3 shows a generic next-generation SONET network.

Over time, Ethernet's current price advantage may diminish if next-generation SONET makes enough of an impact on the equipment marketplace. Next-generation SONET/SDH products are classified as newly designed systems combining legacy voice features with enhancements such as statistical multiplexing.

Key: The primary objective of next-generation SONET development is to improve the scalability and flexibility of SONET while maintaining its fault-tolerant features and resiliency. Many of the carriers surveyed preferred PoS and next-generation SONET over other technologies, due to the familiarity and ease of integrating the new services into their existing infrastructure.

Figure 5-3
Next-generation
SONET network
topology

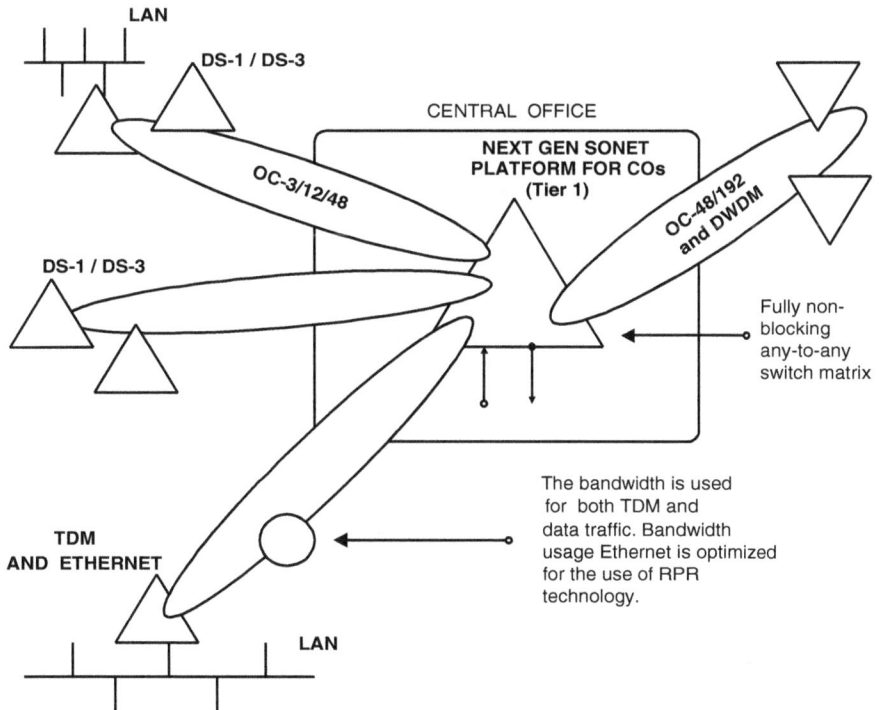

LAN
DS-1 / DS-3
CENTRAL OFFICE
NEXT GEN SONET
PLATFORM FOR COs
(Tier 1)
OC-3/12/48
OC-48/192 and DWDM
DS-1 / DS-3
Fully non-blocking any-to-any switch matrix
The bandwidth is used for both TDM and data traffic. Bandwidth usage Ethernet is optimized for the use of RPR technology.
TDM AND ETHERNET
LAN

The stage is set for a better way to keep the good features and replace the dysfunctional features of traditional SONET. Next-generation SONET vendors realize the need to respect the incumbent telcos' massive investments while at the same time offering them a way to create new services for market differentiation purposes.

As mentioned in Chapter 2, "Metro Area Ethernet," the RBOCs have a massive, sunk investment of several trillion dollars in legacy infrastructure, which won't go away quickly. Do the carriers and hardware vendors consider next-generation SONET a competing or complementary product to their embedded base? That's still a tricky question. Some carriers are deploying new, Ethernet-based products on a standalone basis. In effect, they are building these networks in parallel to their SONET networks. Other carriers are seeking ways to transport Ethernet over their embedded base of SONET infrastructure. Some carriers are following both approaches.

GigE will emerge as a complementary technology to next-generation SONET, not a replacement.

Key: The Yankee Group believes that Ethernet will prevail as an access and transport technology by 2010 due to its ubiquity and simplicity. SONET may be pretty ubiquitous in telco networks, but in the end the hundreds of millions of Ethernet desktops worldwide will have a more dominant impact on the adoption of GigE in the metro.

In the near term (2002 to 2005), SONET and SONET variations will continue to dominate the architectures of the incumbent carriers and even many greenfield operators simply because the carriers have spent a considerable amount of time and money adopting SONET. The universal adoption of GigE in the metro area is not a given until the problems inherent in the current versions of Ethernet are addressed (such as support for QoS, voice, latency, redundancy, and so on).

Next-generation SONET needs to offer the following:

- **Multiprotocol support**   Many service providers offer both voice and data services. Supporting existing and new protocols at the same time by using one platform helps eliminate unnecessary or duplicate network layers. This lowers operational and capital costs for service providers. Enterprise customers benefit by leveraging a single connection for both voice (TDM) and data traffic.

- **High availability**   New solutions must meet the 99.999 percent availability level expected of carrier-class platforms. Next-generation solutions must also meet the enterprise manager and service provider's requirements for restoration levels equivalent to circuit-based switching options (that is, sub 50-ms restoration time in the event of a link failure).

- **Improved utilization** Next-generation SONET solutions improve SONET network utilization by using bandwidth for protection across many wavelengths instead of just one. Another way to improve utilization is to pack SONET channels tighter using other technologies, such as ATM.

- **Ease of migration** New platforms must fit with existing systems and services, and provide a smooth migration path for future networking conditions.

- **New services enabler** The threat of service commoditization is eased by new platforms that support both new and enhanced services, such as offering different levels of service or restorability on the same physical network.

5.4.1 The Options

One difficulty in evaluating next-generation SONET solutions is deciding between the many options carriers have for upgrading their preexisting SONET networks and services. For example, a carrier may choose to deploy a new edge aggregator platform that handles IP, ATM, and SONET to get better network efficiencies and economies of scale. Along with streamlining operations, this also addresses the underlying issue of insufficient network capacity. The same problem can be solved a different way entirely by choosing to deploy DWDM systems to add more wavelengths (and capacity) instead of a next-generation SONET solution.

There isn't a single right answer. The answer can change depending on many variables, including the installed network, management preferences, political pressure from legacy vendors, the type of service provider, their financial state, and the future direction for their services.

Key: The most serious competitor to next-generation SONET is GigE (optical Ethernet). GigE has caught the attention of customers because it offers bandwidth in flexible increments (which are often as small as 1 Mbps) at very competitive rates compared to tariffed SONET services.

One broad category could be called *multiservice SONET*. These solutions typically consolidate cross-connect, ADM, SONET (TDM), and data services (including IP, GigE, ESCON, *Fibre Connection* (FICON), frame relay, and ATM) into one NE (that is, one *box*). SONET can be implemented in either a full or lite version to provide QoS or circuit multiplexing. Vendors do this in at least three different ways:

- **SONET-based solutions that translate everything into SONET and transport the traffic in SONET format** SONET's advantages (guaranteed latency, path protection, and QoS) are retained while providing a migration to packet-based networks. This is really the meaning of next-generation SONET.

- **Interworking platforms that translate all traffic into a single protocol for switching within the platform, but different line cards that can support different services or protocols (such as Nortel's OPTera 5200)** These multiservice platforms use service- or protocol-specific switching fabrics to handle different traffic streams. TDM traffic may be supported either natively or through encapsulation into other protocols, depending on the vendor platform.

- **Multiservice SONET solutions (*multiservice provisioning platforms* [MSPPs]), which generally lower a service provider's costs by consolidating many functions into a single platform** These equipment vendors claim to increase the average utilization of existing networks simply by packing more traffic onto the network. This is accomplished by offering customers connections in VT1.5 increments—service that's much more granular than what today's SONET service provides. These solutions usually sit at either the metro access or edge and may also include some metro core capabilities, especially if DWDM is integrated into the solution.

The disparity of all the multiservice SONET solutions available makes them difficult for carriers to evaluate. Multiservice SONET platforms vary by vendor. All vendors support different degrees or alternate forms of QoS, protection switching, DWDM functionality, and IP transport. The downside? Technical and management complexity associated with provisioning traffic across four separate network layers is inherent in these solutions, providing an implementation challenge (or perhaps a nightmare) for providers. These solutions offer an incremental step in the direction toward integrated networks. For providers with data-only services or business units, optical Ethernet solutions make more sense.

The issue for next-generation SONET is primarily that some overhead exists—in the form of equipment complexity, which is associated with maintaining compatibility with the existing network. However, equipment complexity is ultimately reduced as it is packed deeper into the silicon, and overall system costs will be significantly lower than those of standard SONET equipment. Cash flow constraints in 2002 favor the adoption of incremental SONET improvements or platforms that integrate easily into existing SONET infrastructures.

Key: Pure IP or optical Ethernet solutions are more likely to be added as separate networks and grown independently (in parallel) from the TDM/SONET network. This is because this strategy will be easier for carriers to implement in the long run. First, it would be less expensive and less complicated to implement than deploying a next-generation or MSPP-based network solution. Second, it would probably be easier to train a technician workforce on a completely new, parallel type of network infrastructure that may ultimately be the only network in place. Third, it would be easier to train younger technicians on a new and improved network architecture than to train older workers to embrace and support a radically newer network technology.

For other next-generation SONET benefits such as faster provisioning times and customer-controlled management and provisioning, the picture is still unclear. Service providers' integration cycle of new technology is generally at least 12 to 18 months. In an *Incumbent Local Exchange Carrier* (ILEC) environment, it sometimes takes even longer with entirely new technology platforms. (Certain network support factions in the incumbent telcos are usually very resistant to change.)

The CLECs are the potential early adopters of these solutions. However, as of the summer of 2002, most CLECs are dead or dying. The remaining CLECs that actually have capital budgets for new platforms see the wide array of choices as too confusing. This leads to longer evaluation and decision cycles in CLEC engineering departments.

5.4.2 Impediments to New Technology Deployments

Carriers are reluctant to adopt new technologies because of the integration with existing back-office systems and OSSs. Even in 2002, if a system isn't TIRKS and OSMINE compliant, the RBOCs won't consider putting it into their networks. There are some exceptions, but they're few and far between.

TIRKS is an ancient Bellcore-originated OSS that is heavily geared toward legacy TDM-based technologies and platforms. OSMINE is a Telcordia-based guideline that the RBOCs use to determine if a vendor's equipment is capable of being installed in a telco *central office* (CO) environment.

Standards are also an issue because most next-generation solutions are based on proprietary vendor mechanisms. However, steps are being taken to standardize these proprietary systems. Several examples are the RPR standard development and the GFP that was standardized in late 2001.

Many next-generation SONET solutions are still in the early stages of development and remain unproven. For the most part, enterprise customers can expect little change in current tariffed services until 2003 at the earliest. In other words, new next-generation services won't be available (tariffed) until this time. This also applies to any major changes to current SONET tariffs and services. If and when incumbent telcos adopt next-generation SONET platforms, new capabilities such as granular bandwidth delivery will probably be offered to compete with current and upcoming GigE services. Pricing will be directly related to the local market's state of competition. The more competitive alternatives there are from different service providers that use different technologies, the better the local prices.

The benefits to incumbents are clear. They can manage the service the same way they manage private lines while reducing their customer's cost of ownership considerably. The customer only needs to provide a native Ethernet interface. There's more good news for the ILECs. Ethernet over super-stable SONET translates into *five-9s* (99.999 percent) reliability and the sort of security associated with TDM links. The best the *Ethernet LECs* (ELECs) can do at this time with their pure-play Ethernet architectures is three or four nines reliability, max.

Will next-generation SONET solutions be widely adopted, or are they simply stopgap solutions in an inevitable migration to a data-only world? The latter will be the more likely scenario by 2010, especially if the RPR standard is approved by mid-2003 and widely adopted.

Other next-generation SONET solutions on the horizon include PON systems that run SONET and *free-space optic* (FSO) systems. Regardless of the approach, service providers need solutions that facilitate the future-proofing of their networks as much as possible.

5.5   SONET-lite

Another version of SONET is under development, which uses less of the SONET overhead as explained in the following sections.

5.5.1   IP over Ethernet over DWDM

When the end-to-end networking requirements of legacy telephone networks are considered, SONET has proven to be ideal for transporting 64 Kbps channels between TDM switches that have 64 Kbps channel-switching granularity. As discussed earlier, in these networks, tributary traffic is packaged into virtual containers of various capacities (such as DS-1 and DS-3), which are then multiplexed into a single broadband SONET stream (OC-12, OC-48, and OC-192).

Key: Since 1999, IP traffic has exceeded 64 Kb telephony traffic on many long-haul routes and this trend will continue down into the metro networks. This trend toward IP-centric networks means that the switching and transmission of all information will be increasingly in the form of IP packets. In the interim, it is necessary to provide protocol conversion between IP- and TDM-centric networks.

By stripping SONET of its virtual container overheads, it has become a good interim solution for transporting packets or cells in ring-based metro networks. This is called *SONET-lite*. MAN technology such as *Fiber Distributed Data Interface II* (FDDI-II) has supported packets or cells and 64 Kb channels simultaneously. There has been a recent resurgence of similar (although proprietary) hybrid-muxing technologies aimed at bridging the gap between the IP- and TDM-centric networks.

5.5.2   Ethernet/IP Transmissions in SONET-lite Systems

Before IP packets can be transmitted, they must first be encoded by a layer 1 protocol and framed by a layer 2 protocol in order to match the characteristics of the transmission media being used (fiber, twisted pair, radio, or wireless). SONET-lite provides these protocol layers, and existing test equipment and network management software can still be utilized.

When today's networking requirements are considered, it's better to select a layer 1 encoding and layer 2 framing protocol that is optimized for both the switching and transmission of packets rather than 64 Kb channels. For example, ATM was originally optimized for the switching and transmission of cells, but this later proved to be too cumbersome and expensive for IP-centric networks. This was due to the need for protocol conversion between IP or Ethernet-framed packets to ATM cells and back again, with little additional benefit. It also added major, interim overhead in the form of the dreaded ATM cell tax. In contrast to this scenario, it's now evident that IEEE 802.3z GigE protocol layers are better suited to the transport of IP packets over optical networks. For example, GigE framing eliminates the need for protocol conversion of tributary 10/100 Mb Ethernet streams.

5.5.3   The Quality of Service (QoS) Issue

On the Internet and in other networks, QoS defines how transmission rates, error rates, and other characteristics (that is, delivery guarantees) can be measured, improved, and, to some extent, guaranteed in advance. QoS is of particular importance for the continuous transmission of high-bandwidth video and multimedia information.

Carrying vast amounts of best-effort IP traffic on TDM-based metro networks designed for high-quality voice is extremely inefficient and definitely not profitable. The challenge for network planners is to design scalable, multiservice networks where the network-offered QoS closely matches the QoS (or lack of QoS) paid for by the subscriber.

Mechanisms to ensure true QoS and transport protection are still being developed. To date, RPR remains the single best hope for the realization of carrier-class Ethernet services. Nortel and Cisco have already deployed proprietary RPR-enabled gear, which demonstrates that the concept works. Ethernet's rise to prominence will certainly go hand in hand with RPR or equivalent schemes as the metro transport market evolves.

A lot of work has been done to improve Ethernet's QoS. Although Ethernet will most likely never completely achieve the same level of QoS as SONET, it is approaching a point where there may not be a great deal of practical difference between SONET and overprovisioned GigE.

Key: Overprovisioning and oversubscription are essentially the same thing. This is when, similar to an airline, a service provider will allot X amount of bandwidth on a data network per customer, knowing that if each customer actually sent a constant stream of X amount of data traffic that someone's transmission would be cheated. The theory behind overprovisioning in data networks stems from the fact that these networks are inherently bursty, so no customer will actually send a continuous stream of data that meets his or her purchased allotment.

5.5.4   New Approaches to QoS

Old-world thinking dictates that it's necessary to introduce circuit layers into a network to achieve end-to-end QoS guarantees. This is not true. The deficiencies of native packet-switched solutions arise not from the absence of additional circuit layers, but from the fact that they cannot manage network resources (such as span bandwidth and buffers) in response to service-level commitments. The solution to this problem is not to add additional circuit or protocol layers, but instead to enhance the existing layers and add network-level intelligence to coordinate network resource management.

To support IP-based services such as videoconferencing, video on demand, interactive gaming, and bundled voice, video, and data services, an Ethernet-based infrastructure must be able to deliver individualized services to customers. Subscriber-based IP QoS and 802.1q VLANs are two technologies that can enhance the bandwidth management capabilities of Ethernet to meet requirements for deploying high-quality, next-generation services such as those listed previously. Service providers can use IP QoS to divide traffic into four distinct classes, each with its own priority-level portion of bandwidth.

5.5.4.1   Class 1   The *low-latency and low-jitter class* provides support for real-time, delay- and jitter-sensitive applications such as *voice over IP* (VoIP). The low-latency class ensures that packets are scheduled to achieve low latency first. Secondly, these packets are shaped to achieve low jitter. This traffic does not require shaping, but it should also be rate limited so that non-low-latency traffic is not completely bandwidth starved. In other words, it should not be allocated most or all of the available bandwidth that exists.

The low-latency and low-jitter classes conform to the *Expedited Forwarding* traffic class that the *Internet Engineering Task Force* (IETF) established in *Request for Comment* (RFC) 2598. Expedited Forwarding requires that the egress rate exceed the ingress rate (similar to a reverse funnel) and works similar to a virtual leased line. The Expedited Forwarding guideline polices and drops packets on network ingress, and shapes traffic on egress to make sure that the connection to the next provider is at the same priority level.

5.5.4.2   Class 2   The *low-loss class* gives packet traffic a higher weight in the scheduler, meaning that more bandwidth will be allocated to this traffic, and this traffic will also be given more buffering. This class is often used for mission-critical applications, and it's equivalent to the IETF Assured Forwarding model (RFC 2597).

5.5.4.3 Class 3 The *best-effort class* means that packets will be scheduled as soon as possible. Traffic in this class gets the leftover bandwidth, and it should be the least expensive. This class is similar to the *available bit rate* (ABR) weighting of traffic in ATM networks.

5.5.4.4 Class 4 Like Ethernet, VLAN capabilities are familiar to enterprise customers and offer a valuable service when extending Ethernet into the access arena. IEEE 802.1q is the VLAN standard, and 802.1q support offers the ability to logically (electronically) segregate subscribers and their traffic on a shared Ethernet interface (pipe). IEEE 802.1q defines two key Ethernet fields that have been added to the Ethernet frame to identify and prioritize traffic: the *VLAN identifier* (VLAN ID) and User Priority Values (802.1p).

The VLAN ID field only allows for 4,096 VLAN IDs. With VLANs, service providers can offer per-subscriber services by assigning each subscriber his or her own unique VLAN tag. Figure 5-4 illustrates how an 802.1q VLAN operates.

Figure 5-4
Operation of 802.1q VLANs

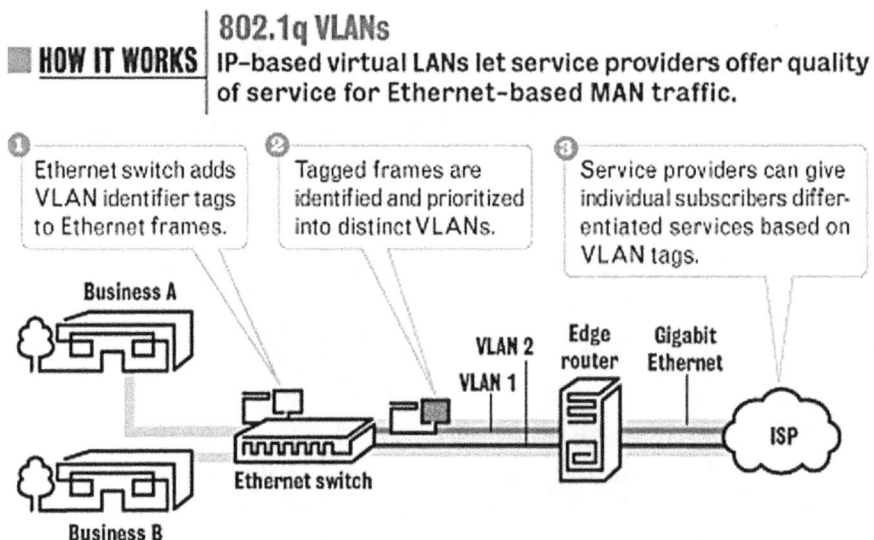

802.1q VLANs

HOW IT WORKS | IP-based virtual LANs let service providers offer quality of service for Ethernet-based MAN traffic.

① Ethernet switch adds VLAN identifier tags to Ethernet frames.

② Tagged frames are identified and prioritized into distinct VLANs.

③ Service providers can give individual subscribers differentiated services based on VLAN tags.

Business A

VLAN 2
VLAN 1

Edge router
Gigabit Ethernet

ISP

Ethernet switch

Business B

5.6   Complementary Protocols and Network Resiliency

Since early 2001, the IETF has been trying to develop standards for MPLS that would extend greater QoS capability into IP networks. The IETF has also been trying to develop *generalized multiprotocol label switching* (GMPLS), formerly known as *multiprotocol lambda switching*, which would extend SONET-like control to DWDM networks. MPLS and GMPLS are generally presented as the QoS answer to IP over DWDM networks. RPR, an alternative to SONET, is also gaining support as yet another service management methodology. Yet GMPLS and RPR are still just baby steps in the search for better methods of management within DWDM and next-generation SONET.

Key: An attractive aspect of GMPLS is its capability to dynamically provision bandwidth (on behalf of) a layer 2 edge device. This makes GMPLS like ATM in functionality and would offer an enhancement to Ethernet that doesn't currently exist. Once GMPLS is deployed, the stage will be set for a network core that dynamically provisions capacity for the access network.

5.6.1   Spanning Tree Protocol

Although Ethernet has enjoyed a major renaissance since around 1999, it has always lacked SONET's reliability and under-50-ms restoration capability. When it comes to resiliency on a metropolitan ring, ELECs and other carriers have few options. They can deploy the spanning tree protocol or the upcoming Fast spanning tree protcool, which will cut restoration time down to 10 seconds or less.

As discussed in Chapter 1, "The Fundamentals of Ethernet Technology," in an Ethernet LAN, computers compete for the ability to use the shared communications path at any given time. If too many computers try to send at the same time, the overall performance of the network can be affected, even to the point of bringing all traffic almost to a halt. To decrease the chances that this will happen, the LAN can be divided into two or more network segments with an Ethernet switch connecting any two segments. Each Ethernet frame goes through the switch before being sent to the intended destination (another LAN segment) or a layer 3 (router) NE. The switch determines whether the message is for a destination within the same segment (as the sender's segment) or for another segment, and forwards it accordingly. The switch does nothing more than look at the destination address in the packet header and, based on its understanding of the two segments (which computers are on which segments), it forwards it to the right path, which really means to the correct outgoing switch port.

Key: The benefit of network segmentation is that the amount of competition for use of the network path is reduced by 50 percent, assuming each segment has the same number of computers. In this scenario, the possibility of the network becoming heavily congested is significantly reduced.

Each Ethernet switch learns which computers are on which segment by sending any first-time message to both segments. This is known as *flooding*. The switch then records the segment from which a computer replied to the message. Then gradually, the switch builds a picture for itself of which computers are on which segments—basically an internal routing table if you will. When subsequent messages are sent, the switch can use this table to determine which segment should be used to send the data. Enabling the Ethernet switch to learn the network through experience is known as *transparent bridging*, meaning that bridging does not require setup by an administrator. In this context, bridging is a logical network activity, not a physical device like in the old days of the 1970s and 1980s.

It is typical to add a second bridge between two LAN segments as a backup in case the primary bridge fails. All segments need to continually understand the topology of the network, even though only one segment is actually forwarding messages. The spanning tree algorithm determines which computer hosts are in which LAN segment, and this data is exchanged using *Bridge Protocol Data Units* (BPDUs). It is broken down into two steps:

Step 1 The spanning tree algorithm determines the best segment a switch can use by evaluating the configuration messages it has received and choosing the best option.

Step 2 Once it selects the primary segment for a particular bridge to send, it compares its choice with possible configuration messages from the nonroot connections it has. If the best option from step 1 isn't better than what it receives from the nonroot connections, it will prune that port.

Key: The software program in an Ethernet switch that enables it to determine how to use bridging between LAN segments is known as the *spanning tree algorithm*. Spanning tree prevents a condition known as a *bridging loop*. A bridging loop describes a network condition where multiple paths linking one segment to another result in an *infinite loop*. The spanning tree algorithm is responsible for defining a bridge that uses only the most efficient path when faced with the option of using multiple paths. If the best path fails, the algorithm recalculates the network and finds the next best route.

The process described above is similar to "discovery" in router-based networks that incur a major link or node failure.

LAN Ethernet switches traditionally employ the spanning tree protcool for redundancy purposes, yielding restoration times of around 30 seconds— far too slow for carrier-grade voice traffic. However, the new IEEE 802.1w *Rapid Spanning Tree Protocol* (RSTP) has cut Ethernet restoration times down to around 10 seconds. Also known as Fast spanning tree protocol, RSTP will cut down on data loss and session timeouts when large Ethernet networks recover after a topology change or device failure.

Key: In real-world scenarios, spanning tree has been shown to provide redundancy on GigE metro networks where two links terminate into one location. When one of the links fails, enterprise data traffic was indeed rerouted to the other link, the only link working. So spanning tree is a viable, complementary protocol for GigE in the metro in terms of providing network resiliency. Yes, the relatively slow restoration time is not ideal and will involve packet loss, but network downtime would be minimal. See Figure 5-5 for an illustration of how spanning tree would provide failover protection.

A final alternative is GigE over DWDM. In this scenario, GigE data streams are multiplexed onto optical wavelengths that are protected by the DWDM system itself, not the Ethernet switches.

In the scenario depicted previously, even though a link failure occurs between locations A and C, A can still communicate with C through its link to B. In this context, B is a communication transit point, not a termination point.

Figure 5-5
Spanning tree
operation in GigE
metro network

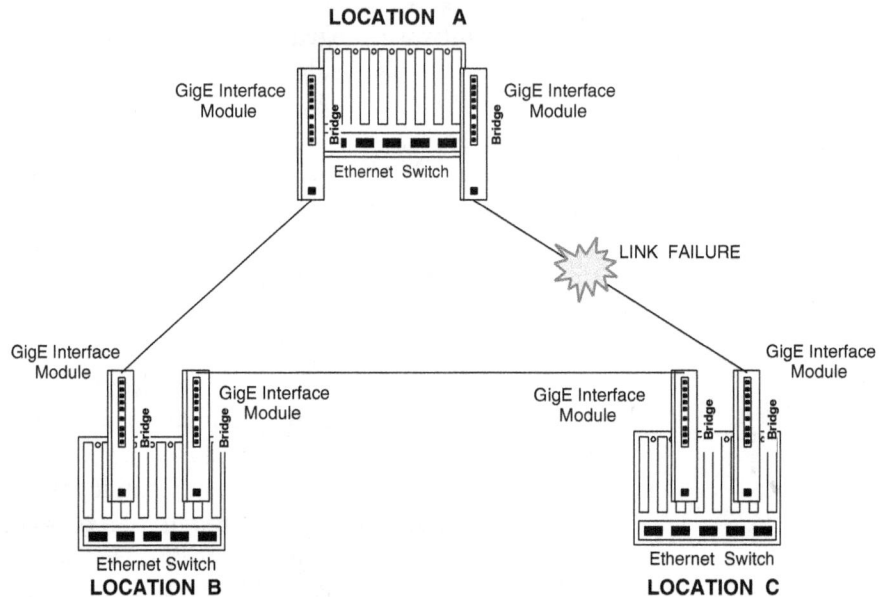

5.6.2 Spatial Reuse Protocol (SRP)

Both RPR and *Dynamic Packet Transport* (DPT) technologies essentially pack twice as much traffic onto fiber rings by using *Spatial Reuse Protocol* (SRP). Cisco's SRP is a proprietary layer 2 protocol that governs topology discovery, protection switching, and bandwidth control instead of dedicating half of the ring's capacity for backup/restoration. This approach isolates faults to a single node (with up to 200 nodes per ring) without the use of TDM.

5.6.3 Dynamic Packet Transport (DPT) Protocol

The concept behind DPT (and RPR) is to take a fiber pair and route traffic on both fibers simultaneously in counter-rotating directions. Unlike SONET (BLSR topology), packet ring technology will use both of the counter-rotating fibers for working traffic. It does not reserve capacity on one entire fiber for protect traffic, so it offers up to twice as much usable network capacity. Because RPR is not TDM-based, traffic is not confined to time slots and can benefit from statistical multiplexing.

DPT has been used by cable modem operators to connect head-end sites to data centers. Now Cisco boasts more than 12,000 installed DPT ports at more than 160 customers, and they expect DPT to migrate to their high-end Ethernet switches, especially the IP-oriented ones.

Key: Compared to an ATM or PoS mesh, DPT gives service providers a way to keep OC-3 port counts low and minimizes the number of IP subnets within a site.

DPT has not taken off in the marketplace because of its pricing and the need to build the entire ring out of high-end Cisco routers, which are very expensive (~$75,000 to $100,000). DPT line cards are used for core architectures based on Cisco 7000 and 12000 series routers, and run at OC-12 and OC-48 speeds.

5.6.4   Multilink Trunking (MLT): 802.3ad

Optical Ethernet is transparent to layer 3 transport and routing protocols, as well as *Domain Name System* (DNS), *Dynamic Host Configuration Protocol* (DHCP), and related tools, so it can work with legacy protocols like *Systems Network Architecture* (SNA) and *Internetwork Packet Exchange* (IPX). Optical Ethernet is scalable from 10 Mbps to 10 Gbps speeds via *multilink trunking* (MLT) (formally known as IEEE 802.3ad Link Aggregation). MLT would be used as a feature in a layer 2 (Ethernet) switch. Multiple GigE data streams are merged together to provide one large multigigabit speed pipe. This pipe would be larger than 1 gigabit, but less than 10 gigabits.

5.6.5   Multiprotocol Label Switching (MPLS)

MPLS is a technology designed to speed up network traffic flow, making it easier to manage. MPLS involves setting up a specific network path (a label-switched path) for a given sequence of packets that are identified by a label put in each packet, thus saving the time needed for a router to look up the address of the next node where the packet should be forwarded.

Key: MPLS can work in a ring, mesh, or point-to-point configuration, and it's called *multiprotocol* because it works with TCP/IP, ATM, and frame relay network protocols.

With reference to the OSI model, MPLS enables most packets to be forwarded at the layer 2 (switching) level rather than at the layer 3 (routing) level. When faced with an option, switching is always preferable to routing because it's faster and more efficient. In addition to moving traffic faster overall, MPLS makes it easy to manage a network for QoS purposes. For these reasons, MPLS is expected to be readily adopted as networks begin to carry many different mixtures of traffic. In addition to moving traffic faster overall, MPLS makes it easy to manage a network for QoS purposes.

MPLS has been specifically designed to compensate for the lack of QoS capabilities in IP networks. It is a network edge protocol that enables all delay-sensitive traffic (voice and video packets) relating to a specific session (or call) to be routed through an IP-based network following a *logical switching path* (LSP). Labels are inserted into packet headers that dictate next hop routing. Network delay is minimized through queuing priority and the establishment of virtual network paths. Preestablished backup LSPs can be used to improve protection-switching capabilities within IP networks. These backup LSPs can also be shared over multiple disjointed paths to reduce protection bandwidth requirements.

Key: To address QoS issues, many optical Ethernet equipment vendors are enhancing their products to support the MPLS protocol. MPLS is being hyped for its ability to scale the Internet through traffic engineering or facilitate IP VPNs by constructing tunnels using its packet labels.

Now MPLS is being considered as a tool to make metro Ethernet as resilient as SONET. In the fall of 2001, the *Metro Ethernet Forum* (MEF) moved to create specifications for the use of MPLS to enable 50 ms restoration of metro Ethernet traffic. This is yet another attempt to make Ethernet carrier class so it can function as a metro transport infrastructure that is more data and packet optimized than SONET. In essence, it is the fast reroute capability of MPLS where the labels correspond to alternative, or detour, label-switched paths, and the fast switching of MPLS that enable 50 ms rerouting, according to the MEF. It may make Ethernet a more appealing transport alternative to SONET for packet data in the metro area, but what about voice traffic? The forum claims this isn't a problem because once MPLS labels and payload data are inserted into an Ethernet frame, the frame can then be prioritized. However, some industry analysts have their doubts about this claim. Time will tell if this works, but if it does work, chalk up another victory for MPLS.

5.6.6   Resilient Packet Ring (RPR)

Another technique to enable carrier-class resiliency in metro area Ethernet networks include the IEEE 802.17 RPR specification, which is scheduled for completion and release in 2003. RPR is an emerging standard focused on facilitating native Ethernet deployments outside of LANs into MAN and WAN topologies where Ethernet traditionally has not been deployed until 1999. RPR strives to create a standard for WAN Ethernet transport that improves service reliability, lowers deployment costs, and facilitates provisioning while adopting Ethernet for ring topologies. Not exclusively designed for Ethernet transmissions, RPR attempts to combine the simplicity, granularity, and high performance of IP packet transmissions with the network protection and survivability of SONET-style optical rings. This isn't to say that Ethernet won't work well with RPR; it just notes that it wasn't designed specifically with Ethernet in mind, but rather packet transport in a generic sense.

RPR is a new layer 2 MAC protocol that combines features of SONET, ATM (statistical muxing), and GigE (simplicity and popularity). As the name indicates, RPR is designed to work in ring topologies and will provide under-50-ms restoration times critical for delay-sensitive (voice) traffic. It will also streamline routing decisions by assuming a ring topology is in place. As a result, routing logic can be simplified to one of three choices: add, drop, and pass through. RPR devices are essentially packet ADMs.

RPR will effectively double the capacity of each metro ring, which is a big plus for service providers who lease their fiber infrastructures. RPR is striving to create a set of standards so that Ethernet can be delivered in its native form without being encapsulated into SONET payloads. It will enable service providers to maximize bandwidth efficiency—for example, not having to provision a SONET OC-48 circuit (of 2.5 Gbps capacity) to transport a 1 Gbps Ethernet signal. RPR should also help carriers facilitate provisioning by not having to deploy a separate, additional layer of SONET rings. RPR standards should make multivendor interoperability easier, which should lead to lower system prices in the long run and help stimulate demand.

As with SONET, packet rings are configured with a working and spare protection ring. Nodes adjacent to each end of a fault can use the ring-wrap technique to route packets in the opposite direction over the protect ring. Protection switching occurs in less than 50 ms. RPR is also being designed to support up to 100 to 200 nodes on one ring, which compares very favorably with the maximum of 16 nodes for a SONET ring if all nodes are TDM and 32 nodes if all nodes are IP. Figure 5-6 illustrates how the ring-wrap process works.

Some equipment vendors already offer proprietary solutions for reliable Ethernet delivery in a WAN environment, but the success of these solutions has been limited. Other equipment vendors also support *prestandard* RPR technology. This includes vendors such as Cisco Systems, Lantern Communications, Luminous Networks, Nortel Networks, or Riverstone Networks.

RPR's objective is to get network restoration time down to 50 ms—the same as SONET. It's easier to implement a fast and robust link-failure recovery mechanism in a ring topology than in a mesh topology. This is because the alternate route is always known in a ring. The IEEE 802.17 Committee does not view RPR as solely Ethernet based, and indeed there is no intent to have an IEEE 802.3 device directly connected to an 802.17 interface.

Although RPR should enable carriers to offer reliable Ethernet service delivery, some major obstacles still exist. Obviously, the most significant impediment is finalizing the parameters of the standard, which is expected in 2003. Another issue that could limit RPR deployments is adoption by service providers. To date, most deployments of proprietary versions of RPR-like solutions have occurred in IP-centric networks, not SONET-based networks.

The implementation of RPR by carriers with a large installed base of SONET equipment could prove challenging. Telco product marketing

Figure 5-6
Ring wrapping in RPR

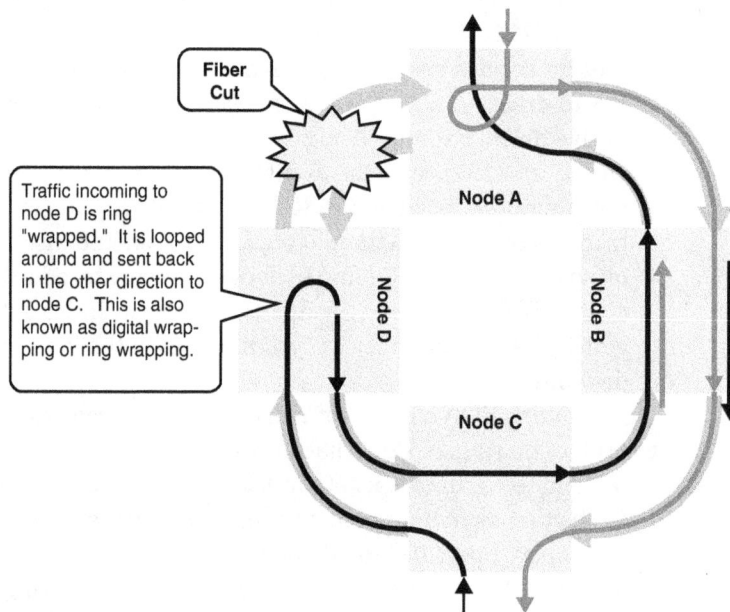

groups need to develop a solid business case that underscores the importance of deploying metro Ethernet products that have an RPR-based infrastructure as their foundations. Even if this means slowly building these networks in parallel to legacy (SONET) networks, in the long run it's still a logical and cost-effective path to a future revenue stream that's certain to grow. The only question is how fast these newer optical Ethernet services will grow. The key aspect of an effort like this is to show (in the business case) that RPR-based Ethernet platforms are more cost-effective than other options such as next-generation SONET.

Key: Most service providers have two distinct transport operations groups: SONET network engineers and IP/Ethernet network engineers. Each group has a separate *capital expenditure* (capex) budget. RPR deployments would require coordination between the two groups, potentially complicating deployments. Sharing a common operating software for the platform could also complicate deployments, management, and billing. In the RBOC world, these groups are actually separated into distinct operational and legal entities via regulatory edicts (the Telecom Act of 1996).

5.6.6.1   RPR Objectives   The current objectives for RPR include using Ethernet framing in SONET-style rings. The goal of RPR is simple: to define a high-performance, high-availability optical transport method suitable for ring-based carrier networks in metro service areas. The RPR standard attempts to define a new MAC to run both Ethernet and SONET on fiber-optic rings, which are the prevalent topologies in metropolitan networks. RPR aims to parcel bandwidth more efficiently than SONET and provide resiliency that's not inherent to Ethernet.

Key: The four primary goals of the RPR Working Group are restoration, resiliency, scalability, and QoS.

As reported by the RPR Alliance, the following objectives comprise a high-level outline of the architecture that the alliance is currently promoting—the features that will make RPR like SONET:

- **Dual counter-rotating ring topology**   In dual-ring topologies, SONET uses only one ring to carry live traffic; the other ring is reserved as a backup. No production traffic is routed across this backup ring, which is a tremendous waste of fiber facilities. To increase fiber utilization, RPR will send traffic over both rings (in opposite directions, of course) during normal operation.

- **A fully distributed access method without a master node**   An RPR ring will continue operating despite the loss of any node.

- **Protection switching in less than 50 ms**   In the event of a fiber break or node failure, RPR will restore service at least as fast as SONET.

- **Destination stripping of unicast traffic**   Unicast traffic is communication between a single sender and a single receiver. In some older packet ring architectures, the source node removes unicast packets after they come all the way around the ring. With RPR technology, destination nodes remove their unicast packets, freeing downstream bandwidth for reuse by other flows. Together with packet multiplexing and counter-rotating rings, destination stripping will more than double RPR's total throughput compared to SONET. See Figure 5-7 for a high-level illustration of the difference between source stripping and destination stripping.

- **Support for multicast traffic**   Multicast traffic is communication between a single sender and multiple receivers. Multicast packets will travel once around the ring to reach every node. By contrast, mesh networks must replicate multicast packets in order to reach all destinations.

- **Support for up to 10 Gbps**   RPR will be fast enough to carry gigabit and 10 GigE traffic, but will also support lower data rates.

- **Support for SONET/SDH (PHY layer), GigE, and 10 GigE (LAN PHY)**   Support for existing PHY layer standards as well as Ethernet up to 10 GigE will enable RPR products to use widely available equipment components.

- **Layer 1 and payload agnosticism**   To be truly universal, the new RPR MAC standard will be completely independent of the PHY layer transport and will not interfere with customer payloads.

- **Plug-and-play support**   New nodes may join the ring without manual configuration—a form of autodiscovery. This will be one of the best advantages of RPR, and compared to SONET provisioning requirements, it will save weeks or months of provisioning time that's normally required when turning up or upgrading a SONET network.

Figure 5-7
The functional
difference between
source stripping and
destination stripping
(Source: "RPR:
Building a Better
Ethernet," Business
Communications
Review
[September 2001])

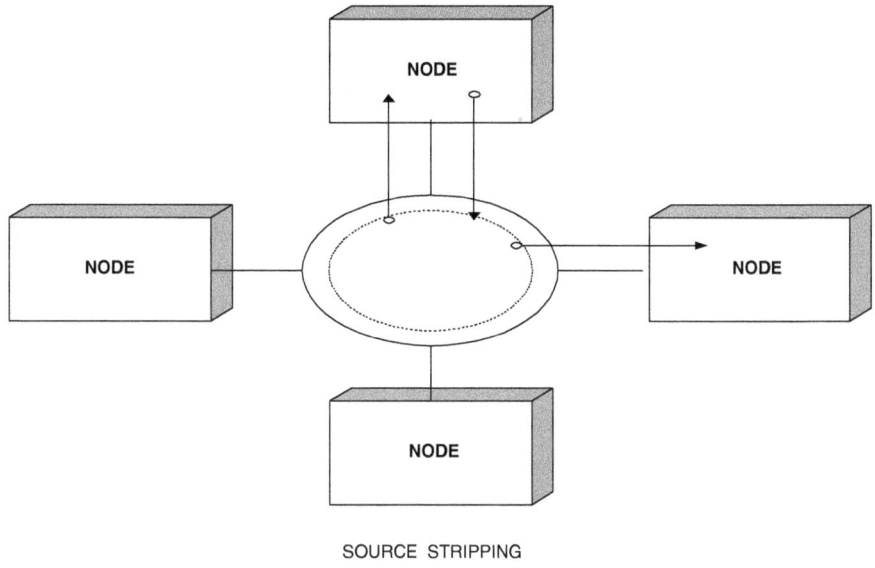

SOURCE STRIPPING

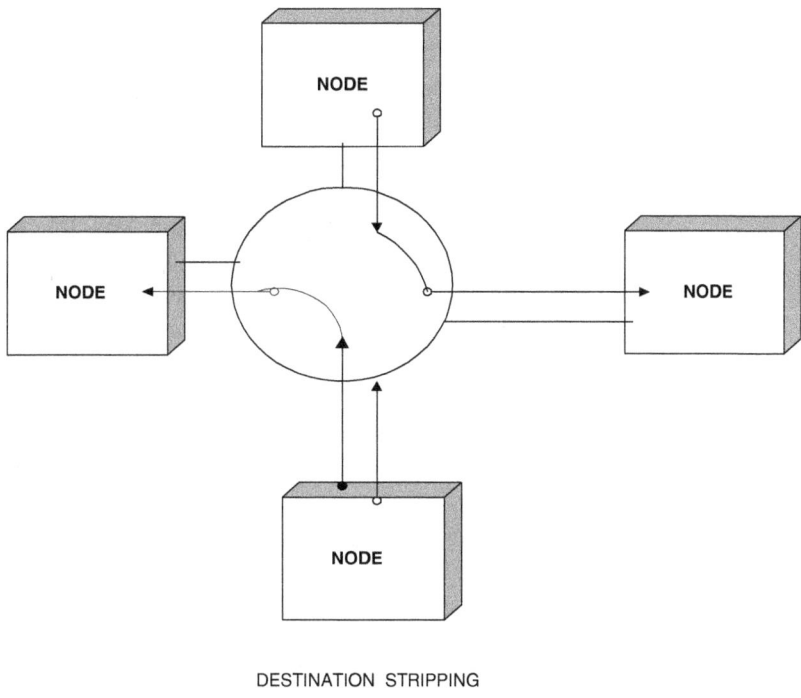

DESTINATION STRIPPING

▪ **Managed objects**   By defining managed objects, the RPR standard will facilitate OSS integration.

▪ **Support for services that require bounded delay and jitter, and guaranteed bandwidth**   RPR will deliver TDM-like QoS. However, the RPR Alliance needs to be very careful with this concept. If RPR moves too far from Ethernet in trying to support TDM, it won't be able to use commodity chipsets, and it may pose more direct competition to SONET than some suppliers would like.

▪ **Dynamic weighted bandwidth distribution**   RPR will allocate bandwidth to competing traffic flows on demand, offering a form of statistical multiplexing. This makes RPR very similar to ATM.

▪ **Support for multiple service types**   RPR will adapt to future requirements.

▪ **Vendor interoperability**   RPR equipment from different vendors will interoperate on the same ring because RPR will be an open standard.

In summary, RPR's primary mission is to make optical rings more efficient for packet traffic, but the standardization effort has attracted some developers who want to see it do more, especially in the realm of QoS and traffic control. Other factions maintain that layer 3 mechanisms such as DiffServ and MPLS can be used to provide these functions, which makes sense.

RPR interoperates with SONET and DWDM, making it suitable for deployment in existing metropolitan optical networks. It can be used to improve the packet-handling capabilities of existing SONET core networks without affecting TDM traffic-handling functions. RPR also supports multiple CoSs.

5.6.6.2 RPR Controversies   Some metro area Ethernet developers are opposed to the new MAC concept that RPR will offer. A vice president from one equipment supplier is quoted as saying that "RPR is as similar to Ethernet as token ring is to Ethernet; that is, they're not similar at all. Ethernet is standard, understood, and based on cheap components that scale fast. RPR will need different components, a different OSS, and so on."

Atrica is one vendor that's taking a different approach with its family of optical Ethernet switches. It uses a 10 Gbps Ethernet MAC over WDM wavelengths and relies on existing standards to make up for Ethernet's shortcomings. For example, DiffServ and 802.1p will let switches manage traffic priorities, while MPLS provides fast recovery from outages.

The engineers at Appian Communications aren't entirely opposed to an RPR MAC, but they are also concerned about its complexity. They believe that matters like fairness and QoS should be handled by existing standards outside of RPR. They believe if the objectives of the RPR working group aren't kept simple, nothing will get accomplished. Good point.

By contrast, Cisco doesn't share Appian or Atrica's concerns. In fact, it already has a new proprietary MAC layer that meets many of the RPR objectives: the SRP. (The SRP is part of Cisco's DPT product line, which has been shipping since late 1999, largely to cable modem service operators. Cisco submitted SRP to the IETF in 1999 as informational RFC 2892 and participates in both the IEEE RPR Working Group [802.17] and the RPR Alliance).

5.6.6.2.1 Packet Priority One of the controversies that the RPR Alliance has been wrestling with is the relationship between packets entering the RPR ring and those already on the ring. Should packets in transit take precedence over entering packets, or should all packets compete equally for bandwidth at every hop? With Cisco's SRP, transit packets take precedence over entering packets, so there's no packet loss on the ring itself. This is different from how Ethernet switches operate. Ordinary Ethernet switches, which lack a cut-through path for transit traffic, exhibit varying packet loss throughout the network as traffic congests at each node. See Figure 5-8.

Figure 5-8

Graphic depiction of transit and entering packets in resilient packet rings

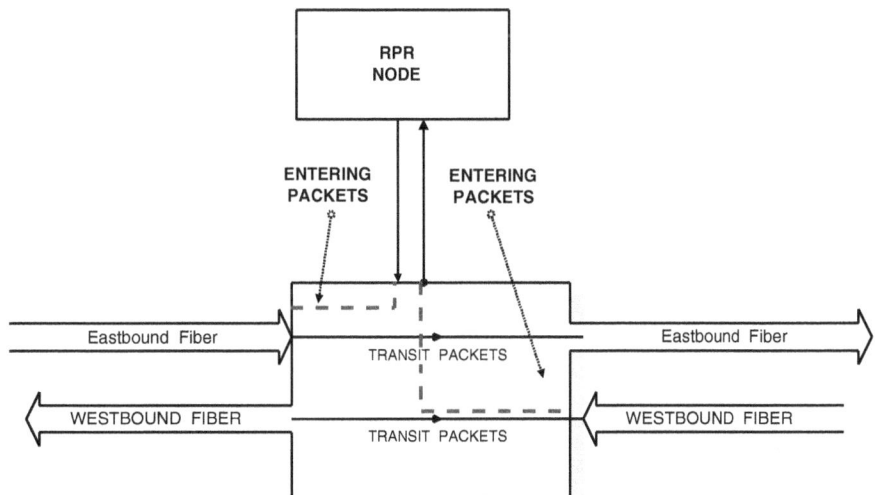

Key: If the 802.17 Working Group decides against transit cut through, then RPR will lose a potential advantage over ordinary Ethernet switching.

5.6.6.3 RPR and TDM: Keep It Simple Folks Another controversial topic is the relationship between RPR and TDM. This controversy involves the definition and scope of RPR, and the migration path from a SONET network to an RPR network.

There are two camps:

- One camp sees RPR as a SONET ADM replacement.
- The other camp sees RPR as simply a high-performance packet MAC.

The rationale of the SONET replacement camp (the first camp) echoes every convergence vendor's slideware. Voice is still a big revenue producer so RPR has to handle voice efficiently. In contrast, the keep-it-simple camp (the second camp) argues that TDM support will be RPR's undoing. A common refrain from the second camp is that if RPR tries to do everything, it won't do enough TDM for the TDM camp and it won't do enough packet for the packet group. They don't want to kill RPR by overloading the working group with a list of objectives that have a huge, varied mix.

Appian Communications offers a hybrid solution to this controversy, a compromise of sorts. Its *Optical Service Activation Platform* (OSAP) dedicates some channels on a SONET ring to packet traffic and leaves other channels for ordinary TDM traffic. This enables OSAPs to share fiber rings with conventional SONET ADMs.

Key: Equipment vendors are likely to ignore any standard that requires massive reengineering of their product lines to support TDM, since today, carriers are mostly content with running their voice circuits over SONET.

Networking history is littered with at least three defunct packet ring standards: FDDI, *Distributed Queue Dual Bus* (DQDB) (802.6), and Token Ring (802.4). To avoid membership in this unfortunate club, the 802.17 Working Group should stick to its main objective: enabling efficient, robust, ring-based metro area data transport. Likewise, if technical complexity makes RPR too expensive, some other flavor of optical Ethernet will capture the market. If the working group tries to be all things to all carriers and equipment makers, standards-based RPR products will never make it past the interoperability demos and the group will have failed.

5.6.6.4 The Competing RPR Proposals In September 2001, two proposed standards competed for acceptance within the RPR Working Group. Cisco's implementation was based on a data-switching architecture where RPR is viewed as a feature of a router or switch. This coalition's approach was to build products to provide optimized SONET circuits to service providers that enable any kind of traffic, including IP, to be carried over a ring.

5.6.6.4.1 The Gandalf Proposal Named after the wizard in J.R.R. Tolkien's *The Lord of the Rings*, the Gandalf proposal was incompatible with Cisco's SRP and would require entirely new silicon, erasing any head-start advantage Cisco would have had with SRP. Cisco conceded on this issue, hoping to get the RPR standard back on track by eliminating the arguments over SRP. Cisco drafted Gandalf with 12 other companies and realized that SRP wasn't going to become the 802.17 standard because at least 26 percent of the vendors would never vote for SRP. The following companies also jumped onto the Cisco bandwagon: Riverstone Networks and several semiconductor companies such as Mindspeed Technologies, Broadcom, and Applied Micro Circuits. The startup Corrigent Systems also jumped on board with support for all or some of Cisco's proposal.

5.6.6.4.2 The Aladdin Proposal Startups, including Luminous and Lantern Communications, formed a coalition to block Cisco's effort, recruiting heavyweight Nortel to anchor their faction. The Aladdin camp preferred the steering method, where every node is notified of a line break and the packets are forwarded away from the break. Gandalf included some provision for steering, but was primarily geared toward wrapping. The Aladdin camp would have preferred to have steering emphasized, and some members preferred to leave out wrapping altogether because it requires implementation at the silicon level. The Cisco camp believed it came up with the preferable option since its proposal would support both methods of protection switching.

Riverstone Networks' concern was that their customers were already asking for RPR. If the draft of the standard wasn't completed on time, the technology could lose credibility and customers would solve the problem using other technology.

The political controversy centered around the fact that Cisco's proposal was based on SRP, a solution that is already shipping on several different versions of Cisco switches and routers with over 13,000 ports deployed. The resentment from the second camp lies in the fact that some companies were worried that if Cisco pushed through its proposal with few changes, it would dominate the market and leave many of the smaller players behind.

Both sides remain optimistic that a single RPR standard can be ratified in 2003 as originally planned. Part of the reason for the optimism is that the Gandalf and Aladdin proposals include some large concessions by both sides. Gandalf and Aladdin presented incompatible approaches, which forced the membership to choose one proposal or the other. The probable outcome will include elements from both Gandalf and Aladdin.

As the 802.17 Working Group mulls over these kinds of debates, simulations will become key tools. The simulations of the competing proposals were one highlight of the January 2002 meeting. This gave the attendees a more concrete view of the differences and an assurance that the algorithms in question would work in line with the standard's goals.

Once the initial draft is completed, the rest is relatively easy. The draft will at least enable equipment manufacturers to build boxes that they know will conform to RPR, even though the final details are not yet hammered out. This is evident in the standards development of 10 GigE, where silicon and systems were shipped in small quantities since late 2001 even though the formal (IEEE 802.3ae) standard wasn't complete until June 2002.

Nortel Networks, Cisco, and Luminous are among the companies with prestandard RPR technology actually installed in live production networks. Many equipment vendors are already shipping prestandard RPR gear. Vendors building RPR-based OEPPs include Luminous Networks and Lantern Communications. Nortel Networks also has an RPR approach through its OPTera Packet Edge System (the OPTera Metro 3500).

5.6.6.5   The RPR Compromise: The Darwin Proposal   The competing Gandalf and Aladdin camps in the RPR standards group crafted a compromise proposal just in time for the group's meeting in late January 2002. The IEEE 802.17 Working Group appeared to be split 50–50 between the two camps, leaving neither side with the 75 percent supermajority needed for ratification. This was the case for months, as Cisco's original proposal stood head to head against the competition.

The compromise, the so-called Darwin proposal, has been posted on the web page of the IEEE 802.17 Working Group. Darwin attempts to bridge the gaps between the competing Gandalf and Aladdin proposals for RPR that were presented at the 802.17 meeting in November 2001. Although it won't necessarily become the 802.17 standard, Darwin presents the compromise that RPR participants felt was inevitable once the Gandalf-Aladdin split formed in 2001.

Darwin forged compromises in the areas of protection and bandwidth management. Darwin would make steering a mandatory part of the RPR standard and offers wrapping as an option. This gives neither faction what it wants, which is usually the case in compromise situations. Gandalf makes wrapping mandatory and offers steering as an option. Conversely, Aladdin makes steering mandatory, but left out wrapping entirely.

On the bandwidth management front, the Darwin proposal assigns unused bandwidth to subscriber connections in a *provisioned unfairness scheme*, giving a larger share of bandwidth to users who paid more for connectivity. The alternative was to distribute the spare bandwidth evenly among all users, as Ethernet does.

The Darwin concept appears to be satisfactory to both camps. A compromise such as Darwin was considered inevitable by most RPR participants, especially since Gandalf and Aladdin already represented softened versions of the opposing stances. Although the RPR tussle is admittedly dramatic, IEEE veterans say the contentious nature of RPR standard development is business as usual for a standard-setting process.

Vendors promote RPR technology as bridging the worlds of SONET and Ethernet and bringing the best features of both technologies to carriers. RPR marries Ethernet's capability to efficiently handle IP data traffic with SONET's under-50-ms protection and the capability to handle voice traffic. By mirroring key SONET features, RPR vendors hope to sell their Ethernet products for both data and voice applications. However, the downside of the RPR approach is that recreating SONET functionality adds significant cost to the equipment, making it significantly more expensive than enterprise Ethernet solutions. For example, recreating SONET's strict timing would require a stratum clock in every CO, and these clocks cost from $50,000 to $100,000 a piece.

The Yankee Group believes that by the end of 2003, all of the major Ethernet switch vendors addressing the carrier market will embrace RPR as a method for offering SONET-like restoration and deterministic QoS to IP networks.

Market acceptance of RPR techniques for handling bread-and-butter circuit services will be slow because RPR equipment vendors will need to extensively prove to incumbent service providers that their approach is a viable alternative to SONET. However, RPR will gain traction with future greenfield players who want to offer Ethernet data services with some level of reliable support for voice traffic. Thus, in the near term, pure RPR products will command a share of greenfield dollars for optical Ethernet infrastructure but few dollars from incumbent carriers. This will only change if and when RPR proves itself useful for transporting all types of traffic. Incumbents may then see the value of RPR and slowly deploy it in their networks.

RPR development is one of the single biggest network technology developments under way in the early twenty-first century. It's the technology that could have the most impact on cementing GigE's place in enterprise and carrier metro networks for many years to come.

CONCLUSION: GIGABIT ETHERNET (GIGE) FOR METRO AREA NETWORKS

Enormous changes took place in telecom technologies, regulatory environments, and multimedia services during the 1990s. The rate of change has been so great that technologies, standards, and products that were initiated in the late 1980s to early 1990s are no longer applicable to the current and emerging market situation. However, this does not mean that only new technologies are applicable to the new market environment. Future metro and full services access networks are likely to be based on Ethernet standards that were first developed in the late 1970s and new *dense wave division multiplexing* (DWDM) technologies that have only recently been brought to market. As a result, 10/100/1000 Ethernet/DWDM access networks will form a new hierarchy that strips away the complexity and cost of existing *Asynchronous Transfer Mode* (ATM) and *Synchronous Optical Network* (SONET) OC-*n* hierarchies, enabling low-cost access to broadband data-centric services upon which all video, audio, telephony, and Internet data services are delivered to homes and businesses. Optical Ethernet may not be the answer to every data communications transport situation, but it will certainly play a major role in the evolution of metro networks in the future PSTN.

APPENDIX A

Helpful Web Sites

www.opticalEthernet.com	Web site of Optical Ethernet information and articles
www.ieee.org	Web site of the IEEE (International Electrical and Electronic Engineers) Association
www.10gea.org	10 Gigabit Ethernet Alliance
www.metroethernetforum.org	Metro Area Ethernet Forum
www.rpralliance.com	Web site of the Resilient Packet Ring (RPR) Alliance
www.sbc.com	SBC Communications
www.lanterncom.com	Lantern Communications
www.cisco.com	Cisco Networks
www.commsdesign.com	Web site of "Communications Design," a web site for telecom industry news, network design, and bulletin boards
www.nortelnetworks.com	Nortel Networks
www.luxn.com	LuxN Communications
www.terabeam.com	Terabeam
www.rbni.com	Redfern Broadband Networks
www.atrica.com	Atrica
www.dynarc.com	Dynarc Communications
www.ieee802.org/17/	RPR Working Group
www.appiancom.com	Appian Communications
www.iec.org	International Engineering Consortium. This site's "On-Line Education/Web ProForum Tutorials" section contains a vast array of excellent tutorials on network and IT-related topics.

www.webopedia.com	Online dictionary and search engine for computer and Internet technology
www.itpapers.com	Web site of white papers and tutorials on IT and network-related topics
www.whatis.com	Online network-related technical dictionary. Also contains many reference documents, white papers, and tutorials.
www.lightreading.com	A premier online newsletter for optical networking topics and news. This site also contains a wide variety of tutorials and white papers.
www.nwfusion.com	Network World Fusion Newsletter. A premier online newsletter for the telecommunications industry related to Network World Magazine.
www.telephony.com	Telephony Magazine
www.Interactiveweek.com	Interactive Week Magazine
www.americasnetwork.com	Web site of America's Network Magazine

GLOSSARY

10BaseT   The IEEE 802.3 specification for Ethernet over UTP.

100BaseFX   100 Mbps Ethernet implementation over fiber.

100BaseTX   100 Mbps Ethernet implementations over CAT 5 and Type 1 cabling.

1000BaseCX   A GigE standard for short-haul shielded balanced copper.

1000BaseLX   A GigE standard for fiber at the 1,300 nm wavelength.

1000BaseSX   A GigE standard for fiber at the 850 nm wavelength.

10 Gb media-independent interface (XGMII)   The interface between the media-dependent and -independent parts of the Ethernet protocol stack.

10 Gigabit Ethernet (10 GigE)   The emerging IEEE standard for Ethernet operation at 10 Gbps.

10 Gigabit Ethernet Alliance (10 GEA)   An organization promoting the rapid deployment of 10 GigE.

802.2   Sets standards at the logical link control sublayer of the data link layer.

802.3   CSMA/CD (Ethernet) standards that apply at the PHY and MAC sublayer.

8B/10B encoding   An encoding scheme at the GigE physical coding sublayer adopted from the FC-1 Fibre Channel specification. It transmits 8 bits as a 10-bit code group.

Active hub   A multiport device that amplifies LAN transmission signals.

Add/drop multiplexer (ADM)   A multiplexer capable of extracting or inserting lower-rate signals from higher-rate multiplexed signals without completely demultiplexing the signal.

American National Standards Institute (ANSI)   The coordinating body for voluntary standards groups within the United States. ANSI is a member of the *International Organization for Standardization* (ISO).

Application program interface (API)   A means of communication between programs to give one program transparent access to another.

Asynchronous Transfer Mode (ATM)   (1) The *Consultative Committee for International Telegraph and Telephone* (CCITT) standard for cell relay wherein information for multiple types of services (such as voice, video, and data) is conveyed in small, fixed-size cells. ATM is a connection-oriented technology used in both LAN and WAN environments. (2) A fast-packet-switching technology providing free allocation of capacity to each channel. The SONET synchronous payload envelope is a variation of ATM. (3) An international *Integrated Service Digital Network* (ISDN) high-speed, high-volume, packet-switching transmission protocol standard. ATM currently accommodates transmission speeds from 64 Kbps to 622 Mbps.

Backbone   (1) The part of a network used as the primary path for transporting traffic between network segments. (2) A high-speed line or series of connections that forms a major pathway within a network.

Backplane   The main bus that carries data within a device.

Bandwidth   (1) The measure of the information capacity of a transmission channel. (2) The difference between the highest and lowest frequencies of a band that can be passed by a transmission medium without undue distortion, such as the *amplitude modulation* (AM) band—535 to 1,705 KHz. (3) Information-carrying capacity of the communication channel. Analog bandwidth is the range of signal frequencies that can be transmitted by a communication channel or network. (4) A term used to indicate the amount of transmission or processing capacity possessed by a system or a specific location in a system (usually a network system).

Bandwidth on demand (BoD)   The dynamic allocation of line capacity to active users.

Bits per second (bps)   (1) The number of bits passing a point every second. The transmission rate for digital information. (2) A measurement of how fast data is moved from one place to another. (For example, a 28.8 modem can move 28,800 bps.)

Border Gateway Protocol (BGP)   The protocol for communications between a router in one autonomous system and routers in another.

Bridge   A device that connects and passes packets between two network segments. Bridges operate at layer 2 of the *Open Systems Interconnect* (OSI) reference model (the data link layer) and are insensitive to upper-layer protocols. A bridge will examine all frames arriving on its ports and will filter or forward a frame depending on the frame's layer 2 destination address.

Bus topology   Linear LAN architecture in which transmissions from network stations propagate the length of the medium and are received by all other stations attached to the medium.

Carrier Sense Multiple Access/Collision Detect (CSMA/CD)   A media access mechanism wherein devices wanting to transmit first check the channel for a carrier. If no carrier is sensed, devices can transmit. If two devices transmit simultaneously, a collision occurs and is detected by all colliding devices, which subsequently delays their retransmissions for some random length of time. CSMA/CD access is used by Ethernet and IEEE 802.3.

Category (CAT)   Often with a number (such as CAT 3) to indicate the grade of UTP wiring.

Category 3 UTP (CAT 3)   An industry standard for an unshielded twisted wire pair capable of supporting voice and low-grade data traffic.

Category 5 UTP (CAT 5)   An industry standard for an unshielded twisted wire pair capable of supporting high-speed data traffic over short distances.

Class of service (CoS)   The categories of traffic used to distinguish between real-time and non-real-time usage as well as between variable and constant bit rates.

Client/server   A distributed system model of computing that brings computing power to the desktop, where users (clients) access resources from servers.

Collapsed backbone   A nondistributed backbone where all network segments are interconnected via an internetworking device. A collapsed backbone may be a virtual network segment existing in a device such as a hub, router, or switch.

Concentrator   (1) A device that serves as a wiring hub in a star-topology network. Concentrator sometimes refers to a device containing multiple modules of network equipment. (2) An FDDI hub.

Connection-oriented network (CON)   Defines one path per logical connection.

Copper Distributed Data Interface (CDDI)   FDDI packets transmitted over CAT 5 UTP cable.

Data Communications Equipment (*Electronic Industries Association* [EIA] expansion) or Data Circuit Terminating Equipment (CCITT expansion) (DCE)   The devices and connections of a communications network that connect the communication circuit with the end device (data terminal equipment). A modem can be considered DCE.

Data link layer   Layer 2 of the OSI reference model. This layer takes a raw transmission facility and transforms it into a channel that appears, to the network layer, to be free of transmission errors. Its main services are addressing, error detection, and flow control.

Differential Services IETF standard (DiffServ) A set of *Internet Engineering Task Force* (IETF) standards designed to provide QoS support in IP networks by providing a means to distinguish among classes of service.

Distance Vector Multicast Routing Protocol (DVMRP) A metrics-based algorithm for routing multicast packets.

Edge device A physical device that is capable of forwarding packets between legacy interworking interfaces (for example, Ethernet, Token Ring, and so on) and ATM interfaces based on data link and network layer information, but does not participate in the running of any network layer routing protocol. An edge device obtains forwarding descriptions using the route distribution protocol.

Enterprise network A geographically dispersed network under the auspices of one organization.

Ethernet (1) A baseband LAN specification invented by Xerox Corporation and developed jointly by Xerox, Intel, and Digital Equipment Corporation. Ethernet networks operate at 10 Mbps using CSMA/CD to run over coaxial cable. Ethernet is similar to a series of standards produced by IEEE referred to as IEEE 802.3. (2) A very common method of networking computers in a LAN. Ethernet will handle about 10,000,000 bps and can be used with almost any kind of computer.

Fast Ethernet The term given to IEEE 802.3u for Ethernet operating at 100 Mbps over CAT 3 or 5 UTP.

FDDI II The proposed ANSI standard to enhance FDDI. FDDI II will provide isochronous transmission for connectionless data circuits and connection-oriented voice and video circuits.

Fiber Distributed Data Interface (FDDI) An emerging high-speed networking standard. The underlying medium is fiber optics, and the topology is a dual-attached, counter-rotating Token Ring. FDDI networks can often be spotted by the orange fiber cable. The FDDI protocol has also been adapted to run over traditional copper wires. (2) An ANSI-defined standard specifying a 100 Mbps token-passing network using fiber-optic cable. Uses a dual-ring architecture to provide redundancy.

Fiber-optic cable A transmission medium that uses glass or plastic fibers rather than copper wire to transport data or voice signals. The signal is imposed on the fiber via pulses (modulation) or light from a laser or *light-emitting diode* (LED). Because of its high bandwidth and lack of susceptibility to interference, fiber-optic cable is used in long-haul or noisy applications.

Fiber optics   A method for the transmission of information (sound, pictures, and data). Light is modulated and transmitted over high purity, hair-thin fibers of glass. The bandwidth capacity of fiber-optic cable is much greater than that of conventional cable or copper wire.

Fibre Channel   Fibre Channel is a high-performance serial link supporting its own as well as higher-level protocols such as FDDI, *Small Computer Systems Interface* (SCSI), HIPPI, and IPI. The fast (up to 1 Gbps) technology can be converted for LAN technology by adding a switch specified in the Fibre Channel standard that handles multipoint addressing.

Full-duplex   Data transmitted in both directions simultaneously. It is available for Ethernet, Fast Ethernet, GigE, and Token Ring. It only supports single stations, not LAN segments.

Gigabit   One billion bits.

Gigabit Ethernet (GigE)   A 1 Gbps standard for Ethernet.

Gigabit Ethernet Alliance (GEA)   An association of GigE manufacturers and suppliers formed for the purpose of promoting GigE technology.

Gigabits per second (Gbps)   One billion bits per second. A measure of transmission speed.

Half-duplex   Data transmitted in either direction, one direction at a time.

Hub   A concentration point for cables. Usually refers to a layer 1 device such as a repeater, concentrator, or MAU.

IEEE 802.1p   An IEEE draft standard that extends the 802.1D Filtering Services concept to provide both prioritized traffic capabilities and support for dynamic multicast group establishment.

IEEE 802.2   IEEE LAN protocol that specifies an implementation of the logical link control sublayer of the link layer. IEEE 802.2 handles errors, framing, flow control, and the layer 3 service interface.

IEEE 802.3ab   IEEE LAN protocol that specifies a UTP cable media implementation of the PHY layer and MAC sublayer of the link layer. IEEE 802.3ab uses a CSMA/CD access at 1,000 Mbps over a variety of physical media.

IEEE 802.3u   IEEE LAN protocol that specifies an implementation of the PHY layer and MAC sublayer of the link layer. IEEE 802.3 uses CSMA/CD access at a variety of speeds over a variety of physical media. One physical variation of IEEE 802.3 (10Base5) is very similar to Ethernet.

IEEE 802.3z   IEEE LAN protocol that specifies a 1,000 Mbps implementation of the PHY layer and MAC sublayer of the link layer.

IEEE 802.3z uses a CSMA/CD access at 1,000 Mbps over a variety of physical media.

IEEE 802.5   IEEE LAN protocol that specifies an implementation of the PHY layer and MAC sublayer of the link layer. IEEE 802.5 uses token-passing access at 4 or 16 Mbps over STP wiring and is very similar to IBM Token Ring.

IEEE 802.6   Standards being developed by IEEE to govern metropolitan area networking.

Institute of Electrical and Electronic Engineers (IEEE)   A professional organization that defines network standards. IEEE LAN standards are the predominant LAN standards today, including protocols similar or virtually equivalent to Ethernet and Token Ring.

Internet   A collection of networks interconnected by a set of routers that enable the networks to function as a single, large virtual network.

Internet Protocol (IP)   A layer 3 (network layer) protocol that contains addressing information and some control information that enables packets to be routed. Documented in RFC 791.

Internetwork Packet Exchange, Network Protocol (IPX)   A layer 3 protocol developed by Novell for NetWare.

Internetworking   A general term used to refer to the industry that has arisen around the problem of connecting networks together. The term can refer to products, procedures, and technologies.

Intranet   A private network that uses Internet software and standards.

LAN segmentation   Dividing LAN bandwidth into multiple independent LANs to improve performance.

Load balancing   In routing, the router's capability to distribute traffic over all its network ports that are the same distance from the destination address. It increases the use of network segments, which increases the effective network bandwidth.

Local area network (LAN)   (1) A network covering a relatively small geographic area (usually not larger than a floor, small building, or campus). Compared to WANs, LANs are usually characterized by relatively high data rates. (2) Network permitting transmission and communication between hardware devices, usually in one building or complex.

MAC layer address   Also called universal address or physical address. A data link layer address associated with a particular network device. Contrasts with a network or protocol address, which is a network layer address.

MAC sublayer   As defined by the IEEE, the lower portion of the OSI reference model data link layer. The MAC sublayer is concerned with media access issues, such as whether token passing or contention will be used.

Management Information Base (MIB)   A database of information on managed objects that can be accessed via network management protocols such as SNMP and CMIP.

Media Access Control (MAC)   IEEE specifications for the lower half of the data link layer (layer 2) that defines topology-dependent access control protocols for IEEE LAN specifications.

Media attachment unit (MAU)   In IEEE 802.3, a device that performs IEEE 802.3 layer 1 functions, including collision detection and the injection of bits onto the network.

Media-independent interface (MII)   The standard in Ethernet devices to transparently interconnect the MAC sublayer and the PHY layer, regardless of the media.

Media interface connector (MIC)   FDDI de facto standard connector.

Megabits per second (Mbps)   A digital transmission speed of millions of bits per second.

Metropolitan area network (MAN)   A data communication network covering the geographic area of a city (generally, larger than a LAN but smaller than a WAN).

Microsegmentation   The division of a network into smaller segments usually with the intention of increasing aggregate bandwidth to devices.

Multimode fiber (MMF)   Fiber-optic cable with a core diameter of 62.5 or 50 microns, in which the signal of light propagates in multiple modes or paths. Since these paths may have varying lengths, a transmitted pulse of light may be received at different times and smeared to the point that pulses may interfere with surrounding pulses. This may cause the signal to be difficult or impossible to receive. This pulse dispersion sometimes limits the distance over which an MMF link can operate supporting the propagation of multiple frequencies of light. Dispersion of light is greater than SMF, so distances are less.

Multiprotocol label switching (MPLS)   A set of IETF standards that are designed to enable packet flows to be switched on the basis of labels instead of the full destination addresses, thereby promoting higher performance and allowing traffic engineering.

Multistation Access Unit (MSAU)   A layer 1 device for the interconnection of Token Ring cables.

Network   A collection of computers and other devices that are able to communicate with each other over some network medium.

Network interface card (NIC)   The circuit board or other hardware that provides the interface between a communicating DTE (for example, a *personal computer* [PC] or server) and the network.

Network layer   Layer 3 of the OSI reference model. Layer 3 is the layer at which routing occurs.

Node   A MAC addressable device joined with others to form a network.

OC-12   The optical carrier level equivalent to SONET STS-3 at 155.52 Mbps.

OC-3   The optical carrier level equivalent to SONET STS-3 at 155.52 Mbps.

OC-4F   The optical carrier level equivalent to SONET STS-4 at 207.36 Mbps.

Open Shortest Path First (OSPF)   A router protocol used within larger autonomous system networks in preference to the Routing Information Protocol (RIP), an older routing protocol that is installed in many of today's corporate networks. Like RIP, OSPF is designated by the Internet Engineering Task Force (IETF) as one of several Interior Gateway Protocols (IGPs). Using OSPF, a host that obtains a change to a routing table or detects a change in the network immediately multicasts the information to all other hosts in the network so that all will have the same routing table information. Unlike the RIP in which the entire routing table is sent, the host using OSPF sends only the part that has changed. With RIP, the routing table is sent to a neighbor host every 30 seconds. OSPF multicasts the updated information only when a change has taken place.

Operation support system (OSS)   The management subsystem for service provider-based networks.

Operations, Administration, Maintenance, and Provisioning (OAM&P)   Tasks performed by the management and administrative systems in a network, especially with reference to public networks.

Optical add/drop multiplexer (OADM)   An ADM used with fiber optics (see ADM).

Optical cable level 3 (OC-3)   Defined standard for the optical equivalent of *synchronous transport signal 3* (STS-3) transmission rate or STS-3c SONET transmission rate. The signal rate for these standards is 155.52 Mbps.

Optical carrier 1 (OC-1) ITU physical standard for optical fiber used in transmission systems operating at 51.84 Mbps.

Optical carrier 3 (OC-3) Optical carrier level 3, SONET rate of 155.52 Mbps; it matches STS-3.

Optical carrier *n* (OC-*n*) Higher SONET level, *n* times 51.84 Mbps.

Packet (1) A logical grouping of information that includes a header and (usually) user data. (2) A continuous sequence of binary digits of information is switched through the network as an integral unit.

Packet buffer Storage area to hold incoming data until the receiving device can process the data.

Packet filtering A second layer of filtering on top of the standard filtering provided by a traditional transparent bridge. It can improve network performance, provide additional security, or logically segment a network to support virtual workgroups.

Physical coding sublayer (PCS) One of the sublayers defined for the Ethernet protocol stack.

Physical layer (PHY) The bottom layer of the OSI and ATM protocol stack, which defines the interface between ATM traffic and the physical media. The PHY consists of two sublayers: the *transmission convergence* (TC) sublayer and the PMD sublayer.

Physical medium dependent (PMD) A sublayer of the physical layer that interfaces directly with the physical medium and performs the most basic bit transmission functions of the network.

Points of presence (POP) The term used by *Internet service providers* (ISPs) to indicate the number of geographical locations from which they provide access to the Internet.

Port (1) The identifier (16-bit unsigned integer) used by Internet transport protocols to distinguish among multiple applications in a single destination host. (2) A connector receptacle in a hub or switch.

Protocol (1) A formal description of a set of rules and conventions that govern how devices on a network exchange information. (2) A set of rules conducting interactions between two or more parties. These rules consist of syntax (header structure), semantics (actions and reactions that are supposed to occur), and timing (relative ordering and direction of states and events). (3) A formal set of rules.

Protocol data unit (PDU) A discrete piece of information such as a frame or a packet in the appropriate format for encapsulation and segmentation in the payload of a cell.

Quality of service (QoS)   The term used for the set of parameters and their values that determine the performance of a given virtual circuit.

Redundancy   The technique of providing duplicate resources for backup purposes.

Repeater (hub)   (1) A device that regenerates and propagates electrical signals between two network segments. (2) A device that restores a degraded digital signal for continued transmission; this is also called a hub or concentrator.

Resource Reservation Protocol (RSVP)   RSVP is a proposed IETF standard that enables Internet applications to request reservation of resources along the path of a data flow so that applications can obtain predictable QoS on an end-to-end basis.

RJ-45   Standard eight-wire connectors for IEEE 802.3 10BaseT networks.

RMON   A standard MIB defined in RFC 1271 to enable networked devices to be remotely monitored.

Router   (1) An OSI layer 3 device that can decide which of several paths network traffic will follow based on some optimality metric. Also called a gateway (although this definition of gateway is becoming increasingly outdated), routers forward packets from one network to another based on network layer information. (2) A dedicated computer hardware and/or software package that manages the connection between two or more networks.

Segment   A bounded section of the network. A segment is usually bounded by bridges, routers, or switches.

Shared Ethernet   Convention CSMA/CD Ethernet configuration to which all stations are attached by a hub and share 10 or 100 Mbps of bandwidth. Only one session can transmit at a time. This is the most popular network type today.

Shielded twisted pair (STP)   Two-pair wire medium used in the transmission of several different protocols. These wires have a layer of shielded insulation to reduce *electromagnetic interference* (EMI).

Simple Network Management Protocol (SNMP)   The Internet network management protocol. SNMP provides a means to monitor and set network configuration and runtime parameters.

Single-mode fiber (SMF)   Also called monomode. SMF has a narrow core (typically 8 to 10 microns in diameter) that enables light to enter only at a single angle. Such fiber has higher bandwidth than MMF, but requires a light source with a narrow spectral width (for example, a laser).

Star topology   A topology where devices are connected to a central point such as a hub.

Switched Ethernet   Configuration supporting an integrated MAC layer bridging capability to provide each port with 10, 100, or 1,000 Mbps of bandwidth. Separate transmissions can occur simultaneously on each port of the switch, and the switch filters traffic based on the destination MAC address.

Switched LAN   Refers to a LAN implemented with packet switches.

Synchronous Digital Hierarchy (SDH)   ITU-TSS international standard for transmission over optical fiber.

Synchronous Optical Network (SONET)   A set of standards for transmitting digital information over optical networks. *Synchronous* indicates that all pieces of the SONET signal can be tied to a single clock. A CCITT standard for synchronous transmission up to multigigabit speeds.

Time division multiplexing (TDM)   A form of transmission in which different flows are combined on the basis of time slots.

Transmission Control Protocol (TCP)   A reliable, full-duplex, connection-oriented, end-to-end transport protocol running on top of Internet protocol.

Transparent bridging   Bridging scheme preferred by Ethernet and IEEE 802.3 networks in which bridges pass packets along one hop at a time based on tables associating end nodes with bridge ports. Transparent bridging is so named because the presence of bridges is transparent to network end nodes.

Transport Control Protocol/Internet Protocol (TCP/IP)   A protocol (set of rules) that provides the reliable transmission of packet data over networks.

Unshielded twisted pair (UTP)   Two- or four-pair wire medium used in the transmission of many different protocols such as Token Ring, 10BaseT, and CDDI.

Virtual LAN (VLAN)   A VLAN segment is a unique broadcast domain. Membership to a VLAN is defined administratively independent of the physical network topology.

Wave division multiplexing (WDM)   A technology that enables multiple wavelengths to be multiplexed over a single strand of fiber. It comes in various forms including coarse, dense, and wide depending on the number of wavelengths involved.

Wide area network (WAN)   (1) A network that encompasses interconnectivity between devices over a wide geographic area. Such networks would require public rights of way and operate over long distances. (2) A network that covers an area larger than a single building or campus.

REFERENCES

Allen, Doug. "Will Gigabit Ethernet WAN Services Make Us Forget About SONET?" *Network Magazine* (September 4, 2001).

————. "10 Gigabit Ethernet Reaches for the Metro," *Network Magazine* (October 2001).

Appian Communications. "Virtual Ethernet Rings: LANs Across the WAN," White Paper, www.appiancom.com, January 2002.

Bellman, Robert. "RPR—Building a Better Ethernet," *Business Communications Review* (September 2001). Used with permission of the publisher, Key3Media, BCR Events, Inc. (www.bcr.com) from *Business Communications Review* (September 2001).

Bogen, Norm. "Metro Optical Networks: Evolution Versus Revolution," Cahners In-Stats/MDR Report, January 2002.

Caisse, Kimberly. "Improving Ethernet," *Network World Magazine* (December 24, 2001).

Clavenna, Scott. "More Generations of SONET to Come," *Business Communications Review* (November 2000).

————. "Endless Ethernet," www.lightreading.com, February 5, 2001.

Conover, Joel. "Networking for the New Generation," *Network Computing Magazine* (June 25, 2001).

Dodd, Annabel. *The Essential Guide to Telecommunications*. Upper Saddle River, NJ: Prentice Hall, 1998.

Duffy, Jim. "Nortel Unveils Core-Specific Multiservice Switch," *Network World Newsletter* (February 28, 2002).

Extreme Networks. "Building New-Generation Metropolitan Area Networks," White Paper, www.extremenetworks.com.

Fabbi, M. "10 Gigabit Ethernet Scales the WAN," Gartner Group Report, February 2, 2001.

Gage, Beth and Scott Williams. "Emerging Technology: Next-Generation SONET: Stopgap or Sure Thing?" *Network Magazine* (August 3, 2001).

Gimpelson, Terri. "Ethernet/SONET Debate Strong as Ever," *Network World Magazine* (June 18, 2001).

————. "MSPPs, Next-Gen SONET Mixing Optical Signals," *Network World Fusion* (October 9, 2001).

Greene, Tim. "Metro Ethernet Could Pay Off for RBOCs," *Network World Fusion* (January 2002).

International Engineering Consortium (IEC). "Optical Ethernet," a tutorial, White Paper, www.iec.org.

———. "The Direction of the Optical Networking Market," WebProForum tutorial, www.iec.org.

———. "Ethernet Passive Optical Networks," WebProForum tutorial, www.iec.org.

———. "Optical Metro Edge," White Paper, www.iec.org.

Judge, Peter. "Ethernet Goes Metropolitan," *IT Week* (December 2001).

Kaplan, Ron. "U.S. Metropolitan Ethernet Services Market Forecast and Analysis, 2001–2006," IDC, 2001.

Kawamoto, Wayne. "New Generation of 10 Gigabit Ethernet Products," *ISP Planet* (May 24, 2001).

Krapf, Eric. "Postponing the Inevitable," *Business Communications Review* (September 2001). Used with permission of the publisher, Key3Media, BCR Events, Inc. (www.bcr.com) from *Business Communications Review* (September 2001).

Lehman Brothers and McKinsey & Co. "The Future of Metropolitan Area Networks," joint study, August 24, 2001.

Livoli, Karen. "Extending Ethernet into the Access Arena," *Network World Magazine* (March 4, 2002).

Lucero, Sam. "Ethernet in the First Mile: A New Paradigm," Cahners In-Stat Group, September 2001.

Matsumoto, Craig. "Compromise Proposed to RPR Standard Group," *EE Times*, published at CommsDesign.com, (January 17, 2002).

———. "Lord of the Packet Rings: Gandalf Battles Aladdin," *EE Times* (January 18, 2002).

Morgan, Jonathan A. "Ring Versus Mesh: The Emerging Metro Topology," www.appiancom.com.

Pappalardo, Denise. "Cogent Grabs PSINet Assets," *Network World Fusion* (March 1, 2002).

Perrin, Sterling. "Optical Ethernet Provisioning Platforms Market Forecast and Analysis, 2000–2005," IDC, 2001.

Reardon, Marguerite. "RPR Meeting Divided," *Network World Newsletter* (September 24, 2001).

Ryan, George P. "Gigabit Ethernet at the Core of the Network," Applied Technologies Group, White Paper, www.itpapers.com.

Ryan, Jerry. "Building 10 Gigabit/DWDM Metro Area Networks," Principal—Applied Technologies Group, White Paper, 2000.

Rybczynski, Tony. "Optical Ethernet—Preparing for the Transition," *Business Communications Review* (October 2001). Used with permission of the publisher, Key3Media, BCR Events, Inc. (www.bcr.com) from *Business Communications Review* (September 2001).

Stasney, Marian. "Gigabit Ethernet in the Metro Market," Yankee Group Report, 2001.

Titch, Steven. "GigE Evolution," *America's Network Magazine* (January 15, 2002).

"An Ethernet Tutorial," www.whatis.com.

INDEX

Symbols

Tellaire, 173–174
Telseon, 169–173
Terabeam, 174–176
TWT, 176–177
Verizon Communications, 177–179
WorldCom, 179–180
XO Communications, 182–183
Yipes Communications, 184–187
Ethernet standard, 8
Ethernet switching, 24
Ethernet versus Fibre Channel, 244
Ethernet versus SONET, 229
  complexity issues, 235
  cost of SONET expansion, 232
  existing systems data service
    evolution, 236
  legacy SONET constraints, 229–230
  long provisioning cycles, 235
  restoration versus efficiency, 233–234
  scalability, 232
Ethernet virtual private lines, GigE
  applications, 81
Ethernet/IP transmissions, SONET-lite, 277
Everest Broadband, Ethernet service
  provider, 165
evolution
  CLECs, 133
  MAN requirements, 41
  PONs, 122
EVPNs (Ethernet Virtual Private
  Networks), 81
existing systems data service evolution,
  SONET issues, 236
expansion of raw fiber optic capacity, 137
experimental Ethernet, 2
explicit rate QoS, packet-optimized
  SONET/SDH, 258
Extreme Networks, 211

F

Fast Ethernet, 12
fast spanning tree, 53

fiber optic, availability and market
  inhibitor, 135
Fibre Channel
  channels, 242
  networks, 243
  versus Ethernet, 244
fields, Ethernet frames, 10–11
first mile security, Ethernet, 120
flooding, 23
flow-through provisioning, 43
Foundry Networks, 212
Frame Relay, 245
frames, Ethernet, 9
FSO (free-space optics), 136, 141
FTTB (Fiber to the Business), 119–120
FTTH (Fiber to the Home), 119
future developments for 10 GigE, 108

G

Gandalf proposal, 295
GBICs (gigabit interface converters), 68
Giant Loop, Ethernet service provider, 166
GigE (Gigabit Ethernet), 12, 68
  802.3z standard, 69
  advantages, 73
  applications, 74, 76
    common threads, 75
    EPL, 80–81
    Ethernet virtual private lines, 81
    EVPNs, 81
    high-speed Internet access, 78
    inter-POP connectivity, 79
    SANs, 79–80
    telemedicine, 83
    TLAN services, 76–77
    virtual Ethernet rings, 82
    virtual private Ethernets, 82
  bandwidth sharing, 69
  comparison to 10 GigE, 95
  disadvantages with SONET, 238–239
  equipment development, 198–200
  LAN migration options, 72–73

ABOUT THE AUTHOR

Paul Bedell obtained an M.S. in Telecommunications Management, with Distinction, from Chicago's DePaul University in 1994. In 1995, Bedell designed and has since taught a course on cellular and wireless telecommunications at DePaul University in the Graduate School of Computer Science, Telecom, and Information Systems.

He began his telecommunications career in the United States Army Signal Corps, serving in (West) Germany from 1985 to 1988, where he worked as a multichannel communications equipment operator at a remote signal site. Upon discharge in 1988, Bedell spent five years working for several Fortune 500 companies as a telecommunications analyst.

From there, he moved to the wireless industry, where he spent five years working for three leading wireless carriers in both the cellular and PCS industries. He spent two-and-a-half years as a network engineer for United States Cellular. During that time, he designed and implemented fixed and interconnection networks in most regions of the United States. He then moved to PrimeCo Personal Communications, where he managed the build out of the prelaunch fixed network for the metro portion of PrimeCo's Chicago market. Bedell then moved to Aerial Communications (which is now known as T-Mobile), where he designed and managed the build out of Aerial's prelaunch *wide area network* (WAN).

Bedell moved to Ameritech's long-distance business unit in 1998, *Ameritech Communications, Inc.* (ACI), where he managed the implementation of ACI's 42-node data network, which supported its new 2,500-mile SONET system that spanned the five Ameritech states. Paul also installed two *voice over IP* (VoIP) networks at ACI before moving into Data Network Product Marketing in August 2000.

Since August 2000, he has been the Associate Director of Product Marketing at SBC Communications, managing a metro area Ethernet product known as *GigaMAN*SM. Bedell managed the expansion of the product into the Pacific Bell and Southwestern Bell territories. GigaMAN was originally introduced by SBC in the Ameritech region in 1999.